MATHEMATICS RESEARCH DEVELOPMENTS

GLOBAL OPTIMIZATION

THEORY, DEVELOPMENTS AND APPLICATIONS

Mathematics Research Developments

Additional books in this series can be found on Nova's website under the Series tab.

Additional e-books in this series can be found on Nova's website under the e-book tab.

Computational Mathematics and Analysis Series

Additional books in this series can be found on Nova's website under the Series tab.

Additional e-books in this series can be found on Nova's website under the e-book tab.

MATHEMATICS RESEARCH DEVELOPMENTS

GLOBAL OPTIMIZATION

THEORY, DEVELOPMENTS AND APPLICATIONS

ANGELIKA MICHALSKI
EDITOR

nova publishers
New York

Library of Congress Cataloging-in-Publication Data

Global optimization : theory, developments and applications / [edited by] Angelika Michalski.
pages cm
Includes bibliographical references and index.
ISBN 978-1-62417-796-5 (hardcover)
1. Mathematical optimization. I. Michalski, Angelika, editor of compilation.
QA402.5.G6195 2013
519.6--dc23
2012050351

Published by Nova Science Publishers, Inc. † New York

Contents

PREFACE

Global optimization is an important area of applied mathematics that has shown fast and growing development during the last few decades, and deals with the optimization of a function or a set of functions according to some criteria. In this book the authors discuss the theory, developments and applications of global optimization including: particle swarm optimization with re-initialization strategies for continuous global optimization; particle swarm global optimization for orbital maneuvers; float-encoded genetic algorithm used for the inversion processing of well-logging dataparticle collision algorithms; classifier-assisted frameworks for computationally expensive optimization problems; and the cutting box strategy as an algorithmic framework for improving metaheuristics for continuous global optimization.

Chapter 1 – Particle Swarm Optimization (PSO) is a population-based algorithm inspired by social behavior of animals such as bird flocking and fish schooling. PSO algorithm can be easily implemented with few parameters involved, and several modifications have been introduced to the original PSO algorithm to discourage premature convergence in global optimization. In this study, the author propose re-initialization strategies for improving the performance of PSO for continuous global optimization. The first re-initialization strategy focuses on individual particle level, where a particle whose pbest fails to improve after certain number of iterations will be reinitialized. The second re-initialization strategy focuses on the population level, where the entire population will be reinitialized. In this case, re-initialization is conducted when the number of pbests having objective function value similar to that of gbest exceeds a certain threshold number. The combination of these two re-initialization strategies has also been studied. The proposed PSO algorithms have been tested using 15 benchmark functions, each with both 10 and 30 variables. The results are compared with the efficient population utilization strategy PSO (EPUS-PSO), comprehensive learning PSO (CLPSO), cooperative PSO (CPSO), unified PSO (UPSO) and unified bare-bones PSO (UBBPSO).

Chapter 2 – The particle swarm optimization algorithm is a population-based stochastic method developed in recent years and successfully applied in several fields of research, ranging from chemical engineering to computer science. It represents a very intuitive methodology for global optimization, inspired by the behavior of bird flocks while searching for food. The particle swarm optimization technique takes advantage of the mechanism of information sharing that affects the overall behavior of a swarm, with the intent of determining the optimal values of the unknown parameters of the problem under consideration. This work describes the application of the method to a variety of space trajectory optimization problems that exhibit multiple local optimal solutions. Specifically, the swarming algorithm is applied to: (i) transfer trajectories between two coplanar circular orbits; (ii) two-impulse transfer between two non-coplanar elliptic orbits; (iii) four-impulse rendezvous trajectory between a pursuer spacecraft and a target space vehicle; (iv) two impulse transfer from a low Earth orbit to a low Moon orbit. Despite its simplicity and intuitiveness, the particle swarm algorithm proves to be quite effective in finding the globally optimal solution to all of the applications considered in the chapter, with great numerical accuracy.

Chapter 3 – A Float-Encoded Genetic Algorithm is presented in this chapter for solving the well-logging inverse problem. The aim of the global inversion of well-logging data is to provide a robust and reliable estimate of petrophysical properties, such as porosity, water saturation, shale volume and mineral content, associated with geological structures. There are two possible ways to solve the interpretation problem. The first is a conventional inversion scheme, separately estimating the unknowns to different depths. In the forward modeling phase of the local inversion procedure the theoretical well-logging data set is calculated by using locally defined probe response functions, which are then fit to real data in order to estimate model parameters to one depth only. This procedure leads to a marginally over-determined inverse problem, which results in relatively poor parameter estimates. A further disadvantage of the latter technique is that some crucial quantities, such as the thickness of layered geological formations, cannot be extracted by inversion because they do not appear explicitly in local response equations. A new inversion methodology introduced by the authors gives much more freedom in choosing the inversion parameters. The so-called interval inversion method inverts all data measured from a greater depth interval in a joint inversion process. By a series expansion-based discretization of the petrophysical model, a highly over-determined inverse problem can be formulated, enabling an estimation of the petrophysical parameters including new unknowns, such as zone parameters and layer thicknesses, more accurately compared to local inversion methods. The authors give further references for several applications of the global inversion method. In this chapter, one synthetic and two field examples are presented to demonstrate the application of the Genetic Algorithm-based inversion method. It is shown that the combination of the new inversion strategy and global optimization tools forms a highly effective and adaptive algorithm for earth scientists who are interested in a more reliable calculation of the reserves of hydrocarbons and other mineral resources.

Chapter 4 – Stochastic optimization methods based on the simulated annealing paradigm have been actively developed in the last thirty years and successfully applied to many complex optimization problems. These methods though very powerful are not free from practical drawbacks, the main one being that performance is too sensitive to the choice of free parameters. Ideally, an optimization algorithm should not rely on user-defined or problem-dependent parameters, except a stopping criterion. One such algorithm, that was recently introduced and applied to real-world problems, is reviewed here. Denoted the "Particle Collision Algorithm" (PCA), the algorithm is loosely inspired by the physics of nuclear particle collision reactions, particularly scattering and absorption. These events promote the exploration of the search space and the exploitation of its most promising areas. The PCA resembles in its structure that of simulated annealing and can also be considered a Metropolis algorithm, as a worse trial solution can be accepted with a certain probability. This acceptance may avoid the convergence to local optima. Since its introduction, some variants of the original algorithm have been created, and it has also been hybridized with other techniques. The PCA and its derivations have been successfully applied to real-world optimization problems, such as a nuclear reactor design optimization problem, radiative transference and pollutant localization inverse problems, chemical equilibrium nonlinear systems, and course-timetabling problems.

Chapter 5 – The modern engineering design optimization process often replaces laboratory experiments with computer simulations. This setup formulates an optimization problem of a black-box function, namely, which has no analytic expression and which is computationally expensive to evaluate. However, computer simulations introduce an additional challenge into the design optimization process: often there will exist candidate designs which cause the simulation to fail, which implies that a large portion of the optimization resources may be wasted, and this can lead to a poor final result. To effectively handle such problems, this chapter describes two optimization frameworks which incorporate a classifier into the search. The classifier's role is to predict which solutions are expected to crash the simulation, and this prediction is then used to bias the search towards solutions for which the simulation is expected to succeed. A baseline framework is described which incorporates a single classifier, followed by a more elaborate framework which selects an optimal type of classifier out of a family candidates. Performance analysis using a representative simulation-driven engineering problem of airfoil shape optimization shows the effectiveness of the proposed frameworks and highlights the merit of incorporating a classifier into the search.

Chapter 6 - In this work, the authors present a new framework to increase effectiveness of metaheuristics in seeking good solutions for the general nonlinear optimization problem, called Cutting Box Strategy (CBS). CBS is based on progressive reduction of the search space through the use of intelligent multi-starts, where solutions already obtained cannot be revisited by the adopted metaheuristic. Computational experiments with the CBS strategy are conducted with a variant of the population-based metaheuristic Differential Evolution to solve 36 test instances. The numerical results show that CBS can substantially increase the quality of the results of a metaheuristic applied for a nonlinear optimization problems.

In: Global Optimization
Editor: Angelika Michalski

ISBN: 978-1-62417-796-5

Chapter 1

PARTICLE SWARM OPTIMIZATION WITH RE-INITIALIZATION STRATEGIES FOR CONTINUOUS GLOBAL OPTIMIZATION

D. D. Kennedy[1], H. Zhang[1], G. P. Rangaiah*[1]
and A. Bonilla-Petriciolet[2]
[1]National University of Singapore,
Department of Chemical and Biomolecular Engineering, Singapore
[2]InstitutoTecnologico de Aguascalientes,
Deparment of Chemical Engineering, Mexico

ABSTRACT

Particle Swarm Optimization (PSO) is a population-based algorithm inspired by social behavior of animals such as bird flocking and fish schooling. PSO algorithm can be easily implemented with few parameters involved, and several modifications have been introduced to the original PSO algorithm to discourage premature convergence in global optimization. In this study, we propose re-initialization strategies for improving the performance of PSO for continuous global optimization. The first re-initialization strategy focuses on individual particle level, where a particle whose *pbest* fails to improve after certain number of iterations will be reinitialized. The second re-initialization strategy

* Corresponding Author: Department of Chemical and Biomolecular Engineering, National University of Singapore, Singapore, 117576, e-mail: chegpr@nus.edu.sg.

focuses on the population level, where the entire population will be reinitialized. In this case, re-initialization is conducted when the number of *pbest*s having objective function value similar to that of *gbest* exceeds a certain threshold number. The combination of these two re-initialization strategies has also been studied. The proposed PSO algorithms have been tested using 15 benchmark functions, each with both 10 and 30 variables. The results are compared with the efficient population utilization strategy PSO (EPUS-PSO), comprehensive learning PSO (CLPSO), cooperative PSO (CPSO), unified PSO (UPSO) and unified bare-bones PSO (UBBPSO).

Keywords: Stochastic Global Optimization, Particle Swarm Optimization, Re-initialization Strategy, Benchmark Functions

NOMENCLATURE

c_1 and c_2 = Acceleration Coefficients of the Particles.

Counter_1 = Counter for the Consecutive Failure of *pbest*.

Counter_2 = Counter for the Number of *pbest*s having the Same Objective Funtion Value as *gbest*.

D = Total Dimension of the Particle.

d = Dimension Counter from 1 to D.

$F(X_i^k)$ = Objective Function Value of ith Particle at k^{th} Iteration.

f_{min} = Global Objective Function Value for Benchmark Functions.

$G_i^{d,k+1}$= Velocity Update of the ith Particle's d^{th} Dimension at k^{th} Iteration Using *gbest* topology.

$gbest^k = (gbest^{1,k}, \ldots, gbest^{D,k})$ = Position of the Global Best Particle at k^{th} Iteration.

$gbest^{d,k}$ = d^{th} Dimension of the Global Best Particle at k^{th} Iteration.

I = Initialization Range for Benchmark Function.

i = Particle Counter from 1 to NP.

it_max = Maximum Number of Iterations.

k = Iteration Counter from 1 to it_max.

$lbest_i^k = (lbest_i^{1,k}, \ldots, lbest_i^{D,k})$ = Position of the ith Local Best Particle at k^{th} Iteration.

$lbest_i^{d,k}$ = d^{th} Dimension of the ith Local Best Particle at k^{th} Iteration.

$L_i^{d,k+1}$ = Velocity Update of the ith Particle's d^{th} Dimension at k^{th} Iteration on *lbest* Topology.

NFE = Number of Function Evaluations.

NP = Number of Particles or Population Size.

np = Number of Problems.

n_i = Number of Neighbors for i^{th} Particle.

$nbest_i^k = (nbest_i^{1,k}, \dots, nbest_i^{D,k})$ = Position of the ith Neighborhood Best Particle at k^{th} Iteration.

$nbest_i^{d,k}$ = d^{th} Dimension of the ith Neighborhood Best Particle at k^{th} Iteration.

Pc = Learning Probability of Particle in CLPSO.

P_i = Particle i^{th} Leader.

Pr = Activating Rate of SRS in EPUS-PSO.

Ps = Learning Probability of Particle in EPUS-PSO.

$pbest_i^k = (pbest_i^{1,k}, \dots, pbest_i^{D,k})$ = Position of the ith Previously Best Particle at k^{th} Iteration.

$pbest_i^{d,k}$ = d^{th} Dimension of the ith Previously Best Particle at k^{th} Iteration.

$pbest_{fi(d)}^d$= d^{th} Dimension of Any Particle *pbest.*

r = U [1, D].

$Rand\_1_i = (Rand\_1_i^1, \dots, Rand\_1_i^D)$ = First Random Number in the Range [0, 1].

$Rand\_2_i = (Rand\_2_i^1, \dots, Rand\_2_i^D)$ = Second Random Number in the Range [0, 1].

S = Search Range for Benchmark Function.

SR = Success Rate.

$Trial_i^d$ = Trial Vector of Particle i in d^{th} Dimension.

U = Unification Factor.

$U[0, 1]$ = Uniform Distribution from Zero to One.

V_{max} = Upper Bound for the Velocity of the Particle.

V_{min} = Lower Bound for the Velocity of the Particle.

$V_i^k = (V_i^{1,k}, \dots, V_i^{D,k})$ = Velocity of the ith particle at k^{th} Iteration.

$V_i^{d,k}$ = Velocity of the dth Dimension of the ith Particle at k^{th} Iteration.

w = Inertia Weight.

x = Unrotated Vector.

xx = Dummy Vector.

X_{max} = Upper Bound for the Position of the Particle.

X_{min} = Lower Bound for the Position of the Particle.

$X_i^k = (X_i^{1,k}, \dots, X_i^{D,k})$ = Position of the ith Particle at k^{th} Iteration.

$X_i^{d,k}$ = d^{th} Dimension of the ith Particle at k^{th} Iteration.

y = Rotated Vector.

Greek Symbols.

$\sigma_i = (\sigma_i^1, \sigma_i^2, \dots, \sigma_i^D)$ = Standard Deviation Vector of ith Particle.

σ_i^d = d^{th} Dimension Standard Deviation of ith Particle.

λ = Number in the Range of [0, 1].

$\mu_i = (\mu_i^1, \mu_i^2, \dots, \mu_i^D)$ = Mean Vector of the ith Particle's *pbest* and *nbest* Vector.

$\mu_i^{d,k}$ = d^{th} Dimension of the Mean of ith Particle's *pbest* and *nbest* at k^{th} Iteration.

1. Introduction

Global optimization is an important area of applied mathematics that has shown fast and growing development during the last few decades. If the objective function is non-convex, optimization problems have many local optima, and this makes finding the global optimum a challenge. In particular, engineering and science applications involve challenging global optimization problems that must be solved efficiently and reliably. Global optimization methods can be broadly divided into two approaches: deterministic and stochastic (or metaheuristics). Stochastic global optimization methods are attractive for solving challenging global optimization problems arising from real-life applications because they are applicable to any problem without the assumptions of continuity and differentiability of the objective function. There are many stochastic global optimization methods which include simulated annealing, genetic algorithms, tabu search, differential evolution, ant colony optimization, and particle swarm optimization (PSO) [1,2].

Specifically, PSO is a popular stochastic global optimization based on simulation of social behavior [3,4]. It makes the population of potential solutions to explore the search space of the optimization problem for the global optimum. The performance of PSO relies on the exchange of information between individuals, called particles, of the population, called swarm. During optimization sequence, each particle adjusts its trajectory towards its own previous best position and towards the best previous position attained by members of its neighborhood. This iterative process is repeated until the termination criterion (e.g., maximum iterations) is met. PSO has gained popularity in the global optimization research community due to its effective performance and ease of implementation; e.g., [5-8].

PSO and its variants have been applied to many benchmark and application problems; e.g., [9-12]. Several studies have tried to enhance the performance of PSO by incorporating mechanisms for proper initialization and dynamic adaptation of PSO parameters, and by the application of new or modified PSO operators that include hybrid methods using local optimization methods and/or other meta-heuristics [6,11-16]. However, even though PSO is usually effective for continuous global optimization [16], the performance of available PSO algorithms may be poor in large-scale optimization problems due to premature convergence and population stagnation caused by the problem complexity and dimensionality. Therefore, it is important to study new algorithm modifications and improvements to develop efficient and robust PSO algorithms suitable for handling complex optimization problems with many decision variables.

In this study, we propose new re-initialization strategies for improving the performance of PSO for continuous global optimization. The first re-initialization strategy focuses on individual particle level, where a particle whose *pbest* fails to improve after certain number of iterations will be reinitialized.

The second re-initialization strategy focuses on the population level, where the entire population will be re-initialized. The combination of these two re-initialization strategies has potential to improve the performance of PSO algorithm.

The performance of the proposed PSO algorithms has been tested using 15 benchmark functions, each with 10 and 30 variables. The results are compared with four variants of PSO algorithm: efficient population utilization strategy PSO (EPUS-PSO) [17], comprehensive learning PSO (CLPSO) [18], cooperative PSO (CPSO) [19], and unified PSO (UPSO) [20]. The comparison suggests that the proposed PSO algorithms perform better for several benchmark functions than the other reported PSO techniques.

The remainder of this paper is organized as follows. Section 2 provides a description of the concepts behind PSO and some of its variants including EPUS-PSO, CLPSO, CPSO and UPSO.

Description of the proposed re-initialization strategies for improving the performance of PSO is given in Section 3. Section 4 presents the results and discusses the performance of the proposed PSO algorithms. Finally, the conclusions of our study are summarized in Section 5.

2. Description and Development of PSO

PSO is a population-based optimization technique proposed by Eberhart and Kennedy [3,4]. It is based on social behavior of animals such as bird flocking and fish schooling. Each particle in the population represents a point in the D-dimensional search space, and serves as a potential solution. The movement of the particle is guided by its velocity which is influenced by the personal best position (*pbest*) found so far by the particle and the neighborhood best position (*nbest*) found so far by neighboring particles. Note that *nbest* can be *gbest* (global best) or *lbest* (local best). The original PSO algorithm updates the particles in the population as follows [3,4]:

$$V_i^{d,k+1} = V_i^{d,k} + c_1 Rand\_1_i^d \left(pbest_i^{d,k} - X_i^{d,k}\right) + c_2 Rand\_2_i^d \left(gbest^d - X_i^{d,k}\right) \tag{1}$$

$$X_i^{d,k+1} = X_i^{d,k} + V_i^{d,k+1} \tag{2}$$

where $Rand\_1_i^d$ and $Rand\_2_i^d$ are two different random numbers in the range [0, 1] and are different for every dimension of the particle, $V_i^{d,k}$ and $X_i^{d,k}$ are the dth element of velocity and position vectors of ith particle, respectively, at k^{th} iteration. In this method, c_1 and c_2 are weights of the stochastic acceleration used to adjust the movement of the particle's dimension towards the corresponding *pbest* and *gbest* dimensions. They are the learning factors towards the best positions, and are commonly termed as the acceleration coefficients.

The upper and lower bounds for V_i^d are V_{max}^d and V_{min}^d respectively. If there is any value of V_i^d lying above V_{max}^d or below V_{min}^d, V_i^d will be reassigned to satisfy these bounds. Similarly, X_i^d is clamped between X_{max}^d and X_{min}^d. After updating each particle and evaluating its objective function, *pbest* of the particle is updated by the following equation:

$$pbest_i^{k+1} = \begin{cases} X_i^{k+1} \; if \; F(X_i^{k+1}) < F(pbest_i^k) \\ pbest_i^k \; otherwise \end{cases} \tag{3}$$

Similarly, *gbest* of the population is updated by the following equation:

$$gbest = \begin{cases} X_i^{k+1} \; if \; F(X_i^{k+1}) < F(gbest) \\ \quad gbest \; otherwise \end{cases} \tag{4}$$

If relevant, *nbest* is also updated similarly. The flowchart of the original PSO algorithm is presented in Figure 1.

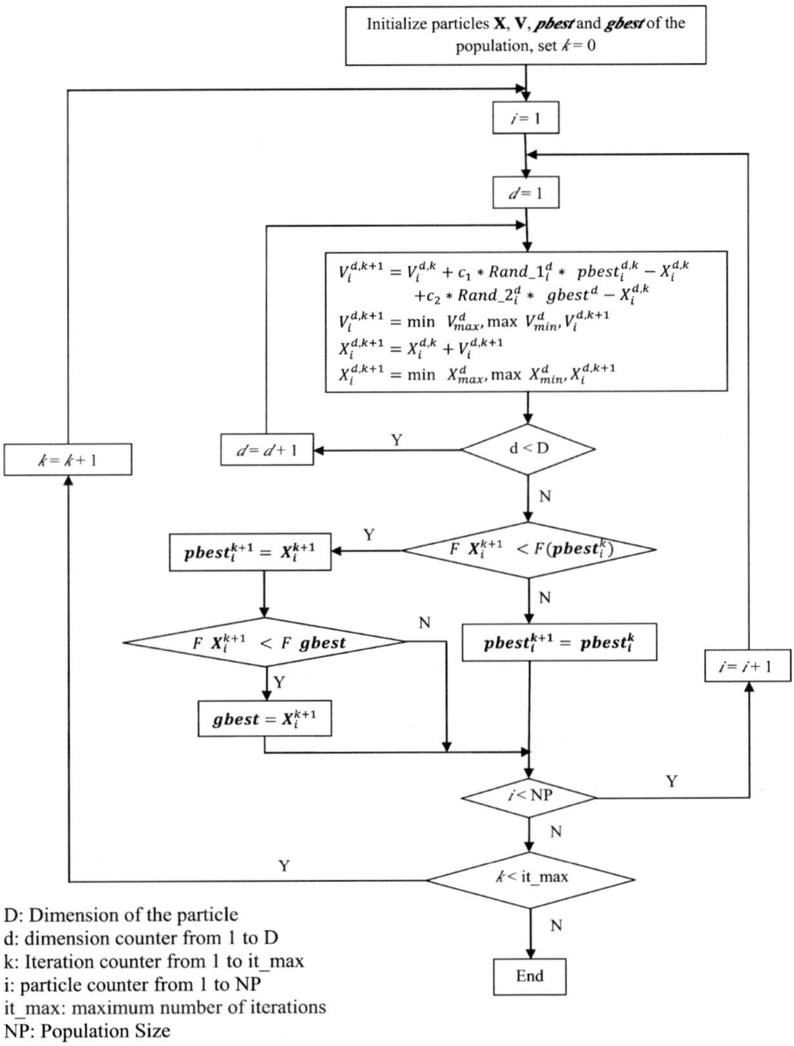

Figure 1. Flowchart of the original PSO algorithm.

Due to its simplicity, PSO has attracted the interest of a number of researchers. Since its introduction, there have been many modifications to the original PSO algorithm. In particular, inertia weight (i.e., coefficient for the first term on the right hand side of equation 1) and a constraint on coefficients c_1 and c_2 in equation 1, were introduced to prevent explosion of particle velocity and to promote the right balance between exploration and exploitation during global optimization. For example, Eberhart and Shi [21] proposed linearly decreasing inertia weight over the course of the search. Using this approach, exploration is more emphasized in the early stage while exploitation is the focus in the later stage of the search. In a later study, Eberhart and Shi [22] designed a nonlinear approach to vary inertia weight using fuzzy methods, and they reported improved performance of the resulting PSO algorithm. Another effective strategy is to implement inertia weight with a random component rather than linearly decreasing inertia weight [23]. Instead of using linearly decreasing inertia weight, Zheng et al. [24] have applied increasing inertia weight which shows reasonable performance.

On the other hand, Kennedy [25] observed that the trajectories of non-stochastic one-dimensional particles contained interesting regularities when c_1+c_2 was between 0.0 and 4.0. Later, Clerc and Kennedy [26] proposed constriction coefficient to control the convergence of the particles, prevent the explosion movement of the particle and eliminate the arbitrary bounds for the velocity. This has also eliminated the guessing for appropriate values of c_1 and c_2. Also, the performance of PSO depends on the population topologies, which can be divided into two main types: static and dynamic topologies. The influence of the social network and the effect of topology depend on the interactions of particles in the population. In fact, several studies have shown the improved performance of PSO using different population topologies [27,28].

Several modified PSO algorithms have also been proposed to improve the efficiency and robustness of the global search. In particular, four recent variants of PSO are discussed as they are used to compare the performance of the proposed PSO algorithms in this study.

Learning strategy based on historical best information of all particlesis employed in the comprehensive learning PSO (CLPSO) [18]. In CLPSO, the velocity update is modified as follows:

$$\begin{cases} V_i^{d,k+1} = w^{k+1}V_i^{d,k} + c_1 Rand\_1_i^d\left(pbest_{fi(d)}^{d,k} - X_i^{d,k}\right) \\ X_i^{d,k+1} = X_i^{d,k} + V_i^{d,k+1} \end{cases} \tag{5}$$

where f_i indicates the specific particle's *pbest* that the particle *i* should follow, which depends on the learning probability, *Pc* that can varyfrom one particle to another. A random number is generated, and, if this random number is larger than Pc_i, the corresponding dimension will learn from its own *pbest*; else, the particle will learn from another particle's pbest. Note that Pc_i is calculated as follows:

$$Pc_i = 0.05 + 0.45 \frac{\left(\exp\left(\frac{10(i-1)}{NP-1}\right)-1\right)}{(exp(10)-1)} \quad (6)$$

A tournament selection is thus introduced to choose another particle's *pbest*. The modification in CLPSO on particle velocity updating results in three main differences from the original PSO:

1) *pbest* of all particles can potentially be used to direct the flying trajectory of the particle;
2) Each dimension of the particle may learn from different *pbest*'s corresponding dimension for a few generations; and
3) There is only one exemplar, as compared to original PSO which has two exemplars (*pbest* and *gbest*).

The modification introduced by CLPSO improves the diversity of the population and leads to better performance of the algorithm for multimodal benchmark functions. In fact, CLPSO shows better performance than eight other PSO algorithms on unimodal benchmark functions, multimodal benchmark functions, rotated multimodal benchmark functions and composition functions [18].

Van den Bergh and Engelbrecht [19] proposed cooperative approach to PSO (CPSO). In this, the population is divided into different swarms or subpopulations searching solely in different dimensions. Instead of one population trying to find the solution for D-dimensional vector, the vector is divided into its individual dimensions so that there will be D swarms, each carrying out optimization in only one dimensional direction. CPSO provides a fine-grained search which is essential for solving high dimension problems. Thus, it has been found in van den Bergh and Engelbrecht [19] that the performance of CPSO, relative to that of the standard PSO, improves withthe problem dimension. A comprehensive explanation on CPSO can be found in van den Bergh and Engelbrecht [19].

On the other hand, Parsopoulos and Vrahatis [20] proposed unified PSO (UPSO) which is based on two different topologies, namely, *lbest* and *gbest* topologies. These topologies are unified to update the velocity of the particle by:

$$G_i^{d,k+1} = w^{k+1}V_i^{d,k} + c_1 Rand\_1_i^d\left(pbest_i^{d,k} - X_i^{d,k}\right) + c_2 Rand\_2_i^d\left(gbest^d - X_i^{d.k}\right) \tag{7}$$

$$w^{k+1}V_i^{d,k} + c_1 Rand\_1_i^d\left(pbest_i^{d,k} - X_i^{d,k}\right) + c_2 Rand\_2_i^d\left(lbest_i^d - X_i^{d,k}\right) \tag{8}$$

$$\begin{cases} V_i^{d,k+1} = uG_i^{d,k+1} + (1-u)L_i^{d,k+1} \\ X_i^{d,k+1} = X_i^{d,k} + V_i^{d,k+1} \end{cases} \tag{9}$$

where G_i^d and L_i^d are the velocity updates of the particle using *gbest* and *lbest* topology respectively. u is a specified number between [0, 1], known as the unification factor. The population will have *lbest* topology if u is zero, and *gbest* topology if u is one. Results have shown that UPSO with $u = 0.5$ performs relatively better than the original PSO. Further discussion on UPSO can be found in Parsopoulos and Vrahatis [20].

Recently, varying population size of the particle swarm has been introduced in Efficient Population Utilization Strategy PSO (EPUS-PSO) [17]. There are three strategies implemented in this algorithm, namely, population manager, solution sharing and searching range sharing (SRS) strategy. In the population manager strategy, population size is varied depending on the improvement of *gbest* after a pre-specified number of iterations. The population size will decrease if there is an improvement to the *gbest* objective function value, or it will increase if *gbest* fails to improve in the pre-specified number of iterations. In the solution sharing, the particle is updated according to the following equation:

$$\begin{cases} V_i^{d,k+1} = \begin{cases} if U[0,1] \geq P_{s_i} \\ wV_i^{d,k} + c_1 Rand\_1_i(pbest_i^{d,k} - X_i^{d,k}) + c_2 Rand\_2_i\left(gbest^d - X_i^{d,k}\right), \\ Otherwise \\ wV_i^{d,k} + c_1 Rand\_1_i(pbest_i^{d,k} - X_i^{d,k}) + c_2 Rand\_2_i(pbest_{ii}^{d,k} - X_i^{d,k}), \end{cases} \\ X_i^{d,k+1} = X_i^{d,k} + V_i^{d,k+1} \end{cases}$$

where

$$P_{s_i} = 0.5\frac{(D-1)\exp\left(\frac{i-1}{NP-1}\right)-1}{2D} \quad (10)$$

where $pbest_{ii}^{d}$ is the *pbest* of another particle other than particle *i*, which has the best objective function value between two randomly chosen particles from the population. Lastly, the SRS strategy is a mutation-like evolutionary strategy, where some particles or dimensions will be perturbed based on SRS activation rate which is calculated from:

$$P_r^k = A_1 + \frac{A_2 k}{it\_max} \quad (11)$$

where A_1 is the initial SRS rate and A_2 is the variation of SRS rate. EPUS-PSO has shown to perform relatively better than the previously mentioned three variants of PSO, namely, UPSO, CPSO and CLPSO [17]. The detailed discussion on EPUS-PSO and its performance can be found in this reference.

The algorithms, UPSO, CPSO, CLPSO and EPUS-PSO have been successfully used for solving continuous global optimization problems, and the reported results of these strategies are used in the present study for comparing the performance of the proposed PSO algorithms.

3. Particle Swarm Optimization with Re-Initialization Strategies

In the present study, two re-initialization strategies are introduced to improve the performance of the original PSO algorithm. The first re-initialization strategy focuses on individual particle level, where a particle whose *pbest* fails to improve after a certain number of iterations is reinitialized and this is termed *selection* strategy. Re-initialization is carried by setting the current particle's position to *gbest* (or *lbest* in one algorithm) and then randomly choosing one of the dimensions of the particle to take a random value from the search range. The reinitialized particle will only be selected and will replace the current particle if it has a better objective function value.

The second re-initialization strategy focuses on the population level, where the entire population is reinitialized, and this is called *escape* strategy. In this case, re-initialization is activated when the number of *pbest* having objective function value same as that of *gbest* exceeds a certain threshold number. Unlike the first re-initialization strategy, the particles are now selected regardless of their new objective function values. Both the *selection* and the *escape* strategies are studied in this manuscript. For consistency throughout this study, the proposed PPSO algorithms will take acceleration coefficients, $c_1 = c_2 = 2$, and the inertia weight will linearly decrease from 0.7 to 0.2 as per equation 12. These parameters have been proven to be competitive [21].

$$w(i) = w_{max} + \frac{(w_{max} - w_{min})\, i}{i\_max} \tag{12}$$

3. 1. PSO with Selection Strategy

This strategy monitors the performance of *pbest* of each and every particle. There is high tendency that the particles are trapped in a certain area when there is no improvement in the *pbest's* objective function values over certain number of iterations. Thus, it is necessary to increase the exploration nature of the search when this occurs. In this proposed algorithm, when the *pbest*'s objective function value of a particle fails to improve after a threshold number of iterations, known as stagnation interval, the particle will undergo a *selection* process. During this process, a dummy particle is introduced. Every dimension of this dummy particle is set to be the same as that of the *gbest* dimension. A random dimension of this dummy particle is then chosen, and the value of this dimension is randomly chosen from the given search range. The dummy particle will only be selected and will replace the current particle if it has a better objective function value. This is a "greedy" approach.

This proposed PSO is known as PSO with *selection* strategy (PPSO-S). Normally if some dimensions in the particle have already obtained values close to the global optimum, there are only a few remaining dimensions, whose values need changing.

This is normally the case in the later part of the search. Thus, the *selection* strategy exploits the *gbest* position and introduces an element of exploration through re-initializing one dimension of *gbest*. Figure 2 shows the pseudo code of PPSO-S. In this case, ***xx*** is a dummy particle which is set to be the same as *gbest*, *dd_1* is a randomly chosen dimension from [1, D] and $Counter\_1_i$ is the counter to keep track on how many times i^{th} particle's *pbest* fails to improve. The *selection* strategy algorithm is shown in the dashed box given in Figure 2.

In this algorithm, *gbest* is not updated continuously but it is updated only after each iteration.

Step 1: Initialization

- Initialize iteration counter as $k = 0$, $Counter\_1_i = 0$
- Initialize *NP* particles $\boldsymbol{X}_i^0, i = 1,2, \dots, NP$ and *NP* velocities $\boldsymbol{V}_i^0, i = 1,2, \dots, NP$
- Evaluate objective function values of *NP* particles
- Initialize *NP pbests* ($\boldsymbol{pbest}_i, i = 1,2, \dots, NP$) and set $\boldsymbol{gbest}$

Step 2: Iteration, k

DO

    IF $Counter\_1_i > Stagnation\ Interval$ **THEN**

        $Counter\_1_i = 0$

        $\boldsymbol{xx} = \boldsymbol{gbest}$

        $dd\_1 = Int(Rand * D + 1)$

        $xx^{dd\_1} = X_{min}^{dd\_1} + Rand * \ X_{max}^{dd\_1} - X_{min}^{dd\_1}$

        IF $F\ \boldsymbol{xx} < F\ \boldsymbol{X}_i^k$ **THEN**

            $\boldsymbol{X}_i^k = \boldsymbol{xx}$

        END IF

    END IF

    IF $F\ \boldsymbol{X}_i^k < F\ \boldsymbol{pbest}_i$ **THEN**

        $Counter\_1_i = 0$

    ELSE

        $Counter\_1_i = Counter\_1_i + 1$

    END IF

    Update $\boldsymbol{pbest}_i$ and $\boldsymbol{gbest}$

    FOR each particle, $i = 1\ to\ NP$

        FOR each dimension, $d = 1\ to\ D$

$$V_i^{d,k+1} = w^{k+1} * V_i^{d,k} + c_1 * Rand\_1_i^d * \ pbest_i^d - X_i^{d,k} \ + c_2 * Rand\_2_i^d * \ gbest^d - X_i^{d,k}$$

$$V_i^{d,k+1} = \min \ V_{max}^d, \max \ V_{min}^d, V_i^{d,k+1}$$

$$X_i^{d,k+1} = X_i^{d,k} + V_i^{d,k+1}$$

$$X_i^{d,k+1} = \min \ X_{max}^d, \max \ X_{min}^d, X_i^{d,k+1}$$

        END FOR

        Evaluate objective function $F\ \boldsymbol{X}_i^{k+1}$

    END FOR

WHILE termination criteria not met

Output: $\boldsymbol{gbest}$

Figure 2. Pseudo code of PPSO-S algorithm.

For the implementation of PPSO-S, the choice for stagnation interval is important. The higher the stagnation interval, the higher is the tolerance for the failure in the improvement of *pbest*. This can result in premature convergence that is hard to "save". However, too small a stagnation interval (e.g. zero) can potentially result in high computational time because every time *pbest* fails to improve, *selection* process is activated. In order to choose the proper stagnation interval, a test has been conducted using 10-D Rastrigin function, which is a multimodal benchmark function with many local minima. The function was solved 10 times using NP of 20, maximum number function evaluations (NFE) of 40000 and different stagnation intervals (1, 5, 10, 20 and 50). Our results indicated that too high a stagnation interval will lead to poor results. Only stagnation intervals of 1, 5 and 10 managed to obtain 100% success rate (SR) in the 10 runs. This is probably because when the stagnation interval is too high, exploration nature of the search is relatively less, and the population tends to be trapped in a local minimum. A successful run in this case was defined as obtaining best objective function value close to 1.0E-6 of the global optimum. The performance of low stagnation interval can be further analyzed by looking at the average NFE of the successful runs, which indicates the efficiency of the algorithm. In general, our preliminary calculations suggested that the smaller the NFE, the more efficient is a particular algorithm. Among 1, 5 and 10 stagnation intervals, the interval of 5 seems to be more efficient since on average, it has the lowest value of NFE, while attaining the same SR as the other two. Also, the average number of selections needed for interval 5 was the lowest compared to other stagnation intervals, which means that the *selection* strategy is more efficient using stagnation interval of 5. Therefore, the *selection* strategy employed in this study uses stagnation interval of 5.

Beside stagnation interval, *selection* strategy needs to decide the number of variables in the dummy particle (***xx***) to change. The dummy particle is set to be the same as *gbest*. In order to study this, two benchmark functions have been chosen, 10-D Rastrigin and 10-D Rotated Ackley (with condition number of one).

In this analysis, 10-D Rastrigin and 10-D Rotated Ackley functions were solved 10 times using NP value of 20, NFE of 40000 and five different settings (1, 2, 3, 4 and 5) for changing the dummy particle's variables. Our results indicated that the better SR is only observed when the algorithm changes one variable of the dummy particle at a time for the *selection* strategy. Changing more than one variable in a dummy particle does not yield significantly better result.

In general, the higher the number of dummy particle's variables is changed, the worse the results obtained are. Also, the average NFE for successful run is very comparable for all the tested parameter settings. Therefore, the *selection* strategy used in this study changes one variable of the dummy particle. This variable *dd_1* is randomly chosen, as shown in the pseudo code of PPSO-S in Figure 2.

3. 2. PSO with Escape Strategy

Similar to the *selection* strategy, *escape* strategy is also a re-initialization strategy, but it looks at the population level rather than the individual particle level. This algorithm is called as proposed PSO with *escape* strategy (PPSO-E). If the number of particles with *pbest* objective function values similar to that of *gbest* exceeds certain threshold number, known as escape rate, re-initialization is activated for the entire population. Escape rate is in terms of percentage of the population size. Escape rate of 25% means that *escape* strategy is triggered if more than 25% of the population *pbest*s' objective function values are equal to that of *gbest*. Variables of all these particles are set to be the same as the prevailing *gbest*'s variables. Then, a randomly selected variable of each of these particles is changed to a randomly selected value from the search range. Unlike *selection* strategy, *escape* strategy is not a "greedy" approach. All the new particles are accepted and will replace the current particles regardless of the new objective function values obtained. On top of that, the *pbest* value will take the new particle value keeping only the *gbest* value unchanged. This strategy is similar to the initialization step of the search. However, in *escape* strategy, the re-initialization of the particle is biased towards the *gbest*. This possibly helps to rescue the population that is trapped in the local minima through the provision of the exploration search in the later part of the search if too many particles have *pbest* similar to the *gbest*. Figure 3 shows the pseudo code of PPSO-E. In this case, *dd_2* is a randomly chosen dimension from [1, D], and *Counter_2* counts the number of *pbest*s which have the same objective function values as that of *gbest* in every iteration. Similar to the *selection* strategy, *gbest* value is not updated continuously but only after each iteration.

The *escape* strategy algorithm is shown in the dashed box in Figure 3. Similar to the *selection* strategy, the parameters in *escape* strategy have to be tuned. The first parameter is the escape rate. To study this, 10-D Rastrigin function was solved 10 times using NP of 20, maximum NFE of 40000 and different escape rates (5, 25, 50, 75 and 90). The results indicated that too high an escape rate results in poor SR value; for example 90% escape rate manages to obtain only 70% SR. Escape rates below 75% successfully obtained 100% SR with escape rate of 25% having the lowest average NFE of successful runs.

Step 1: Initialization

- Initialize iteration counter as $k = 0$,
- Initialize *NP* particles $\boldsymbol{X}_i^0, i = 1,2,\ldots,NP$ and NP velocities $\boldsymbol{V}_i^0, i = 1,2,\ldots,NP$
- Evaluate objective function values of *NP* particles
- Initialize *NP pbests* ($\boldsymbol{pbest}_i, i = 1,2,\ldots,NP$) and set $\boldsymbol{gbest}$

Step 2: Iteration, k

DO

Update $\boldsymbol{pbest}_i$ and $\boldsymbol{gbest}$

Counter_2 = 0

FOR each particle, i = *1 to NP*

IF F $\boldsymbol{pbest}_i$ = F $\boldsymbol{gbest}$ **THEN**

Counter_2 = *Counter_2 + 1*

END IF

END FOR

IF *Counter_2* > *Escape Rate* **THEN**

FOR each particle, i = *1 to NP*

$dd\_2 = Int(Rand * D + 1)$

$\boldsymbol{X}_i^k = \boldsymbol{pbest}_i = \boldsymbol{gbest}$

$X_i^{dd\_2,k} = pbest_i^{dd\_2,k} = X_{min}^{dd\_2} + Rand_i * \; X_{max}^{dd\_2} - X_{min}^{dd\_2}$

END FOR

END IF

FOR each particle, i = *1 to NP*

FOR each dimension, d = *1 to D*

$V_i^{d,k+1} = w^{k+1} * V_i^{d,k} + c_1 * Rand\_1_i^d * \; pbest_i^d - X_i^{d,k} \; + c_2 * Rand\_2_i^d * \; gbest^d - X_i^{d,k}$

$V_i^{d,k+1} = \min \; V_{max}^d, \max \; V_{min}^d, V_i^{d,k+1}$

$X_i^{d,k+1} = X_i^{d,k} + V_i^{d,k+1}$

$X_i^{d,k+1} = \min \; X_{max}^d, \max \; X_{min}^d, X_i^{d,k+1}$

END FOR

Evaluate objective function, F $\boldsymbol{X}_i^{k+1}$

END FOR

WHILE termination criteria not met

Output: $\boldsymbol{gbest}$

Figure 3. Pseudo code of PPSO-E algorithm.

High escape rate impedes the execution of escape strategy, which increases the ease for the population to be trapped in the local minima. In addition, 25% escape rate converges relatively faster than the others, and hence this escape rate has been chosen. As mentioned previously, the number of dummy particle's variables to be changed for re-initialization is essential for the selection strategy. Similarly, for the escape strategy, this parameter has been studied using 10-D Rastrigin and 10-D Rotated Ackley functions. Both were solved 10 times using NP of 20 and 40000 NFE. Our results suggest that changing one variable of the new particle at a time is the most efficient approach, as it obtained 100% SR and the lowest NFE with the minimum number of escapes on average. In view of these results, changing only one variable of the new particle is employed in the escape strategy.

3. 3. PSO with Both Selection and Escape Strategies

In order to take advantage of the strengths of *selection* and *escape* strategies, both are implemented together and this algorithm is termed as proposed PSO with *Selection* and *Escape* strategies (PPSO-SE). The pseudo code for PPSO-SE with the *gbest* topology is presented in Figure 4.

Step 1: Initialization

- Initialize iteration counter as $k = 0$
- Initialize NP particles $\boldsymbol{X}_i^0, i = 1,2,\dots,NP$ and NP velocities $\boldsymbol{V}_i^0, i = 1,2,\dots,NP$
- Evaluate objective function values of NP particles
- Initialize NP *pbests* $(\boldsymbol{pbest}_i, i = 1,2,\dots,NP)$ and set $\boldsymbol{gbest}$

Step 2: Iteration, k

DO
  IF $Counter\_1_i > Stagnation\ Interval$ **THEN**
    $Counter\_1_i = 0$
    $\boldsymbol{xx} = \boldsymbol{gbest}$
    $xx^{dd\_1} = X_{min}^{dd\_1} + Rand * X_{max}^{dd\_1} - X_{min}^{dd\_1}$
    IF $F\ \boldsymbol{xx} < F\ \boldsymbol{X}_i^k$ **THEN**
      $\boldsymbol{X}_i^k = \boldsymbol{xx}$
    END IF
  END IF
  IF $F\ \boldsymbol{X}_i^k < F\ \boldsymbol{pbest}_i$ **THEN**
    $Counter\_1_i = 0$
  ELSE
    $Counter\_1_i = Counter\_1_i + 1$
  END IF
  Update $\boldsymbol{pbest}_i$ and $\boldsymbol{gbest}$
  Counter_2 = 0
  FOR each particle, *i = 1 to NP*
    IF $F\ \boldsymbol{pbest}_i = F\ \boldsymbol{gbest}$ **THEN**
      Counter_2 = *Counter_2* + 1
    END IF
  END FOR
  IF *Counter_2* > *Escape Rate* (0.25*NP) **THEN**
    FOR each particle, *i = 1 to NP*
      $\boldsymbol{X}_i^k = \boldsymbol{pbest}_i = \boldsymbol{gbest}$
      $X_i^{dd\ 2,k} = pbest_i^{dd\ 2,k} = X_{min}^{dd\_2} + Rand_i * X_{max}^{dd\_2} - X_{min}^{dd\_2}$
    END FOR
  END IF
  FOR each particle, *i = 1 to NP*
    FOR each dimension, *d = 1 to D*
      $V_i^{d,k+1} = w^{k+1} * V_i^{d,k} + c_1 * Rand\_1_i^d * pbest_i^d - X_i^{d,k} +$
        $c_2 * Rand\_2_i^d * gbest^d - X_i^{d,k}$
      $V_i^{d,k+1} = \min\ V_{max}^d, \max\ V_{min}^d, V_i^{d,k+1}$
      $X_i^{d,k+1} = X_i^{d,k} + V_i^{d,k+1}$
      $X_i^{d,k+1} = \min\ X_{max}^d, \max\ X_{min}^d, X_i^{d,k+1}$
    END FOR
    Evaluate objective function at the resulting $\boldsymbol{X}_i^{k+1}, F\ \boldsymbol{X}_i^{k+1}$
  END FOR
WHILE termination criteria not met
Output: $\boldsymbol{gbest}$

Figure 4. Pseudo code of PPSO-SE algorithm which combines both PPSO-S and PPSO-E algorithms.

The parameters used in this combined strategy are the optimum parameters discussed in Sections 3.1 and 3.2 for *selection* and *escape* strategies. The effect of topology on the PPSO-SE was also studied later, using *gbest* and *lbest* (of two consecutive neighbours) topologies known as PPSO-*gbest*-SE and PPSO-*lbest*-SE respectively.

Note that some times the variable and the velocity of a particle can go beyond the bounds during the search. In fact, there are several ways of handling this situation. In this study, two different approaches are studied, as shown below:

$$\textit{Approach1 1}\begin{cases} V_i^d = \min\left(V_{max}^d, \max\left(V_{min}^d, V_i^d\right)\right) \\ X_i^d = \min\left(X_{max}^d, \max\left(X_{min}^d, X_i^d\right)\right) \end{cases} \quad (13)$$

$$\textit{Approach 2:}\begin{cases} V_i^d = V_{min}^d + Rand\ (V_{max}^d - V_{min}^d) \\ X_i^d = X_{min}^d + Rand(X_{max}^d - X_{min}^d) \end{cases} \quad (14)$$

Thus, approaches 1 and 2 assign, respectively, the nearest boundary value and a random value between the bounds, to the out-of-bound dimension. In order to choose one of these two approaches, a 10-D Rosenbrock function was solved 10 times by PPSO-*gbest*-SE using NP of 20 and NFE of 40000. These results indicated that there is no significant difference in the results of both these approaches. Therefore, approach 1 is implemented in PPSO strategies as shown by the pseudo code presented in Figures 2 to 4.

4. Results and Discussion

4. 1. Description of Benchmark Problems

For studying the performance of different variants of PPSO, 15 benchmark functions consisting of unimodal, multimodal and rotated versions of unimodal and multidomal functions are used. The PPSO algorithms are run in Microsoft Excel 2007. The details of the benchmark functions, the search range and initialization range can be found in Table 1. In this table, S stands for search range, I stands for initialization range, and f_{min} is the objective function value at the global optimum. The first three functions in Table 1 are unimodal and relatively easy to solve, except for *f3-Rosenbrock*. It has a narrow valley near its optimum, thus making it hard for the algorithm to find the global optimum [18]. The remaining benchmark functions are multimodal. *f4-Rastrigin* is a complex multimodal function with numerous local minima making it hard for the algorithm to reach the global optimum as particles can be easily trapped in the local minima. *f6-Ackley* has a narrow global optimum basin and many local minima surrounding it [18].

The rotated benchmark functions 9 to 15 in Table 1 are formed using an orthogonal matrix *M*. The original vector **x** is left multiplied by the orthogonal matrix *M* to get a new rotated vector **y**. The following equations illustrate how a rotated vector is generated:

$$M = \begin{bmatrix} m_{11} & \cdots & m_{1D} \\ \vdots & \ddots & \vdots \\ m_{D1} & \cdots & m_{DD} \end{bmatrix}$$

$$x = [x_1, x_2, \dots, x_D]^T \tag{15}$$

$$y = [y_1, y_2, \dots, y_D]^T$$

$$y = M\,x$$

where

$$y_1 = m_{i1}x_1 + m_{iD}x_2 + \cdots + m_{iD}x_D,\ i = 1,2,\dots,D \tag{16}$$

In this way, the rotated benchmark functions 9 to 15 can be simply generated by replacing the *x* vector with the *y* vector. They are more challenging as a change of one dimension in vector *x* will result in a change of all dimensions in vector *y*. Thus, solving the rotated function using just D one-dimensional searches is not possible. Note that the orthogonal matrix does not change the shape of the function. In this study, the orthogonal matrix is generated using Salomon's method [29] with condition number of one for all the seven rotated functions. The orthogonal matrix M for 10-D and 30-D functions is given in Appendix A.

The PPSO algorithms are evaluated based on both reliability and computational efficiency. Reliability of is indicated by the SR obtained by the algorithm while the computational efficiency is indirectly assessed using the median convergence profile of the algorithm. The results for the benchmark functions are presented using the mean, standard deviation and the SR from all the runs. Fair comparison will be done using the same NFE which is a good indicator of the computational time needed for every run when the function evaluation is computationally expensive.

Table 1. Benchmark Functions used for testing the performance of PSO Algorithms

Benchmark Function	D	S I	Global Optimum	f_{min}
f1 - Sphere $f_1(x) = \sum_{i=1}^{D} x_i^2$	10, 30	$(-100, 100)^D$ $(-100, 75)^D$	[0,..,0]	0
f2 - Quadric $f_2(x) = \sum_{i=1}^{D}\left(\sum_{j=1}^{i} x_i^2\right)^2$	10, 30	$(-100, 100)^D$ $(-100, 75)^D$	[0,..,0]	0
f3 –Rosenbrock $f_3(x) = \sum_{j=1}^{D-1}[100(x_i^2 - x_{i-1})^2 + (x_i - 1)^2]$	10, 30	$(-2.048, 2.048)^D$ $(-2.048, 2.048)^D$	[1,..,1]	0
f4 –Rastrigin $f_4(x) = \sum_{i=1}^{D} x_i^2 - 10cos(2\pi x_i) + 10$	10, 30	$(-5.12, 5.12)^D$ $(-5.12, 3.5)^D$	[0,..,0]	0
f5 - Non-continuous Rastrigin $f_5(x) = \sum_{i=1}^{D} y_i^2 - 10cos(2\pi y_i) + 10$ $y_i \begin{cases} x_i & \|x_i\| < 1/2 \\ \frac{round(2x_i)}{2} & \|x_i\| \geq 1/2 \end{cases}$	10, 30	$(-5.12, 5.12)^D$ $(-5.12, 3.5)^D$	[0,..,0]	0
f6 – Ackley $f_6(x) =$ $= -20\exp\left(-0.2\sqrt{\frac{1}{D}\sum_{i=1}^{D} x_i^2}\right)$ $-\exp\left(\frac{1}{D}\sum_{i=1}^{D}\cos(2\pi x_i)\right) + 20 + e$	10, 30	$(-32.768, 32.768$ $(-32.768, 25)^D$	[0,..,0]	0
f7 - Weierstrass	10, 30	$(-0.5, 0.5)^D$ $(-0.5, 0.35)^D$	[0,..,0]	0

Benchmark Function	D	S I	Global Optimum	f_{min}
$f_7(x) = \sum_{i=1}^{D}\left(\sum_{k=0}^{kmax}\left[a^k cos\left(2\pi b^k(x_i + 0.5)\right)\right]\right) - N\sum_{k=0}^{kmax}\left[a^k cos(2\pi b^k * 0.5)\right]$ $a = 0.5$, $b = 3$, $kmax$=20				
f8 - Penalized Function 2 $f_8(x) = 0.1\left\{ sin^2(3\pi x_1) + \sum_{i=1}^{D-1}(x_i-1)^2\{1+sin^2(3\pi x_{i+1})\} + (x_n-1)^2\{1+sin^2(2\pi x_D)\}\right\} + \sum_{i=1}^{D} u(x_i, 5, 100, 4)$ $u(x,a,k,m) = \begin{cases} k.(x-a)^m, x > a \\ 0, -a > x > a \\ k.(-x-a)^m, x < -a \end{cases}$	10, 30	$(-50, 50)^D$ $(-50, 50)^D$	[1,..,1]	0
f9 - R.Quadric $f_9(x) = Rotatedf_2$ (x)	10, 30	$(-100, 100)^D$ $(-100, 75)^D$	[0,..,0]	0
f10 – R.Rosenbrock $f_{10}(x) = Rotatedf_3(x)$	10, 30	$(-2.048, 2.048)^D$ $(-2.048, 2.048)^D$	[1,..,1]	0
f11 – R.Rastrigin $f_{11}(x) = Rotatedf_4$(x)	10, 30	$(-5.12, 5.12)^D$ $(-5.12, 3.5)^D$	[0,..,0]	0
f12 – R.Non-Continuous Rastrigin $f_{12}(x) = Rotatedf_5(x)$	10, 30	$(-5.12, 5.12)^D$ $(-5.12, 3.5)^D$	[0,..,0]	0
f13 – R.Ackley $f_{13}(x) = Rotatedf_6(x)$	10, 30	$(-32.768, 32.768$ $(-32.768, 25)^D$	[0,..,0]	0
f14 – R.Weierstrass $f_{14}(x) = Rotatedf_7(x)$	10, 30	$(-0.5, 0.5)^D$ $(-0.5, 0.35)^D$	[0,..,0]	0
f15 – R.Penalized Function 2 $f_{15}(x) = Rotatedf_8(x)$	10, 30	$(-50, 50)^D$ $(-50, 50)^D$	[1,..,1]	0

Each benchmark function is solved 30 times with NP of 20, NFE of 40,000 and 150,000 for 10-D and 30-D functions respectively. In this case, a successful run is defined as any *gbest* objective function value obtained that is within 1.0E-6 from the global optimum.

The results and discussion are divided into three sub-sections. The first sub-section covers comparison of the different variants of PPSO, the second sub-section compares the best performing PPSO algorithms with the recent PSO algorithms:s EPUS-PSO, CLPSO, CPSO and UPSO, and the third sub-sections covers the comparison with the unified bare-bones PSO, UBBPSO of Zhang et al. [12]. The results in bold in the tables indicate the best or comparable results for that particular benchmark function among the different algorithms.

4. 2. Performance of Proposed PSO

There are four different PPSOs tested in this case. The first two, PPSO-*gbest*-SE and PPSO-*lbest*-SE, are to study the effect of topology. The last two algorithms, PPSO-*gbest*-S and PPSO-*gbest*-E, are used to study the individual effect of the re-initialization strategies. All the four algorithms are tested using 30-D benchmark functions, and the results are presented in Table 2. A first look at the performance of different re-initialization strategies shows that PPSO-*gbest*-E obtained better or comparable mean and SR for all benchmark functions tested when compared to PPSO-*gbest*-S, except for *f2-Quadric* and *f9-R.Quadric.* This can possibly be explained by the nature of these two functions which have a flat global optimum region. Re-initializing the entire population using *escape* strategy around this flat global region may affect the exploitation nature of the search. This leads to poorer performance of PPSO-*gbest*-E than PPSO-*gbest*-S for *f2-Quadric* and *f9-R.Quadric*. The combination of *escape* and *selection* strategies in general improves the algorithm's performance when compared to the algorithm with individual re-initialization strategy (Table 2).

PPSO-*gbest*-SE gives best results for *f2-Quadric* and *f9-R.Quadric* where PPSO-*gbest*-E has shown poor results for these two functions. However, there are some exceptions such as *f3-Rosenbrock*, where PPSO-*gbest*-E outperforms PPSO-*gbest*-SE. Nevertheless, overall PPSO-*gbest*-SE has a better SR than PPSO-*gbest*-E, and so PPSO-*gbest*-SE is used to compare with other recent PSO techniques in the next sub-sections.

Comparing the results of PPSO-SE based on different topologies, it can be observed that PPSO-*gbest*-SE is performing relatively better for unimodal functions than PPSO-*lbest*-SE (Table 2).

Table 2. Performance Results of Four Variants of PPSOs on 30-D Benchmark Functions

Benchmark Function	PPSO-*gbest*-SE		PPSO-*lbest*-SE		PPSO-*gbest*-S		PPSO-*gbest*-E	
	Mean (Std)	SR (%)	Mean (Std)	SR (%)	Mean (Std)	SR (%)	Mean (Std)	SR (%)
f1- Sphere.	9.04E-45 (5.98E-45)	100	9.48E-26 (2.36E-25)	100	1.64E-41 (9.00E-41)	100	7.02E-45 (3.68E-45)	100
f2- Quadric.	1.12E-04 (1.28E-04)	0	7.03E+00 (3.00E+00)	0	2.11E-04 (2.48E-04)	0	3.87E+03 (4.53E+03)	0
f3- Rosenbrock.	2.44E+01 (2.51E+01)	0	2.87E+01 (2.15E+01)	0	1.95E+01 (1.87E+01)	0	3.21E+0 (2.70E+01)	0
f4- Rastrigin.	0.00E+00 (0.00E+00)	100	1.22E-14 (1.49E-14)	100	0.00E+00 (0.00E+00)	100	0.00E+00 (0.00E+00)	100
f5- Non- Continuous Rastrigin.	0.00E+00 (0.00E+00)	100	1.67E-14 (1.52E-14)	100	0.00E+00 (0.00E+00)	100	0.00E+00 (0.00E+00)	100
f6- Ackley.	4.21E-14 (8.45E-15)	100	1.29E-12 (1.82E-12)	100	1.11E-14 (3.49E-15)	100	7.19E-45 (1.94E-45)	100
f7- Weierstrass.	0.00E+00 (0.00E+00)	100	3.93E-06 (2.15E-05)	97	8.53E-03 (1.25E-02)	3	0.00E+00 (0.00E+00)	100
f8- Penalized Function 2.	1.35E-32 (0.00E+00)	100	5.09E-15 (4.13E-15)	100	3.66E-03 (6.67E-03)	73	1.35E-32 (0.00E+00)	100
f9- R.Quadric.	1.21E-03 (1.71E-03)	7	7.98E+00 (3.54E+00)	0	1.66E-03 (2.93E-03)	3	6.13E+03 (5.76E+03)	0
f10- R.Rosenbrock.	2.70E+01 (1.42E+01)	0	3.09E+01 (1.74E+01)	0	2.79E+01 (1.66E+01)	0	4.68E+01 (4.03E+01)	0
f11- R.Rastrigin.	1.38E+02 (2.77E+01)	0	1.29E+02 (3.11E+01)	0	1.40E+02 (2.81E+01)	0	9.64E+01 (3.88E+01)	0
f12- R.Non-Continuous Rastrigin.	2.04E+02 (3.55E+01)	0	1.88E+02 (2.95E+01)	0	2.03E+02 (4.04E+01)	0	1.56E+02 (3.15E+01)	0
f13- R.Ackley.	3.05E+00 (7.66E-01)	0	2.26E+00 (1.06E+00)	3	3.42E+00 (6.01E-01)	0	2.09E+01 (1.78E-01)	0
f14- R.Weiertrass.	2.24E+01 (4.68E+00)	0	2.00E+01 (3.81E+00)	0	2.29E+01 (3.58E+00)	0	19.7E+01 (19.7E-01)	0
f15- R.Penalized Fun. 2.	2.17E+01 (1.50E+01)	0	2.33E+01 (1.51E+01)	0	2.92E+01 (1.83E+01)	0	1.80E+01 (2.47E+01)	0

This is because *gbest* topology has a greater extent of exploitation search nature than *lbest* topology.

As unimodal function has no local optimum, it is not possible for the population to be trapped in local minima, which implies that exploration is needed relatively less than exploitation search. Moreover, PPSO-*gbest*-SE is more robust in terms of SR than PPSO-*lbest*-SE for *f7-Weierstrass*. For rotated functions *f10* to *f15*, all the four PPSO variants have very low SR. In general, PPSO-*gbest*-SE is more competitive than PPSO-*lbest*-SE.

In order to study the effect of *selection* and *escape* strategies in different benchmark functions, the average numbers of escapes and selections have been captured and summarized in Table 3 using PPSO-*gbest*-SE algorithm. The average number of escapes varies across different functions, and in general rotated functions activate higher number of escapes in a run. The highest escape number is activated by *f12-R.Non-Continuous Rastrigin*, with an average number of escapes of 785 per run. Unimodal benchmark functions tend to have low escape number, even when it is rotated, for example *f2-Quadric* and *f9-R.Quadric*. However, *f1-Sphere* is not consistent with this trend, which may be due to its fast convergence rate, as shown in Figure 5. So in this case, *escape* strategy is activated because most of the particles are in the global optimum region and not because the population is trapped. This suggests that *escape* strategy is activated unnecessarily for some benchmark functions. Despite this weakness of *escape* strategy, it does not compromise on the convergence rate of the PPSO-*gbest*-SE algorithm.

According to Table 2, both PPSO-*gbest*-SE and PPSO-*gbest*-S give comparable results showing results for *f1-Sphere*, despite the presence of *escape* strategy in PPSO-*gbest*-SE. Lastly, high number of escapes does not guarantee high SR value. An example is *f12-R.Non-Continuous Rastrigin,* which has the highest number of escapes in a run but yet PPSO-*gbest*-SE obtains 0% SR.

The average number of selections is more than that of escapes (Table 3), as *selection* strategy is on individual particle level, and the number reflects the number of times the particles in the population undergo *selection* process. Table 3 gives both the total average number of selections and the average acceptance of the dummy particle generated through the *selection* process. Acceptance in this case is based on the objective function value as explained above.

The benchmark function with the highest number of selections, 18108, is *f2-Quadric* with acceptance rate of 13%. The *selection* strategy works most efficiently for *f4-Rastrigin* with the highest acceptance rate of 52%. This is also supported by the results obtained in Table 2, where PPSO-*gbest*-SE has 100% SR for *f4-Rastrigin*. However, higher acceptance value may not necessarily translate into good results; for example, for *f12-R.Non-Continuous Rastrigin*, number of selections is very high with acceptance rate of 28%, but only 0% SR by PPSO-*gbest*-SE (Table 2). From this discussion, it can be concluded that *escape* strategy cannot differentiate between convergence and trapping in local minima when it is activated. As for *selection* strategy, high acceptance rate may not necessarily translate into high SR.

Table 3. Average Number of Escapes and Selections for 30-D Benchmark Functions for 30 Runs obtained from PPSO-*gbest*-SE Algorithm

Benchmark Function	Ave. Number Of Escapes / Run	Ave. Number of Selections / Run		
		Total	Acceptance	% of Acceptance
f1-Sphere.	13	12631	794	6
f2-Quadric.	0	18108	2364	13
f3-Rosenbrock.	12	13128	1507	11
f4-Rastrigin.	5	8989	4654	52
f5-Non-Continuous Rastrigin.	14	9549	4881	51
f6-Ackley.	31	12491	855	7
f7-Weierstrass.	7	9000	2006	22
f8-Penalized Function 2.	64	12364	1092	9
f9-R.Quadric.	0	17825	2362	13
f10-R.Rosenbrock.	14	12680	1327	10
f11-R.Rastrigin.	176	10968	1024	9
f12-R.Non-Continuous Rastrigin.	785	7731	2142	28
f13-R.Ackley.	144	11242	1056	9
f14-R.Weiertrass.	81	13510	684	5
f15-R.Penalized Fun. 2.	19	14434	1241	9

4.3. Comparison of PPSO with Recent PSO Algorithms

From the previous discussion, PPSO-*gbest*-SE and PPSO-*lbest*-SE have been found to be better among the variants of PPSO. These two algorithms are compared with the recent PSO algorithms, namely, EPUS-PSO, CLPSO, CPSO and UPSO. This comparison is based on both 10-D and 30-D of 15 benchmark functions in Table 1. Tables 4 and 5 present the results for 10-D and 30-D, respectively, by PPSO-*gbest*-SE and PPSO-*lbest*-SE along with the results for EPUS-PSO, CLPSO, CPSO and UPSO taken from Hsieh et al. [17]. In these tables, SR values for PPSOs are also given but not for other algorithms as such results were not reported in Hsieh et al. [17]. In general, since the global minimum value of benchmark functions tested is zero, mean value greater than zero indicates that SR of the method may be much less than 100%; for example, SR is only 7% when the mean is 1.21E-3 for *f9* function by PPSO-*gbest*-SE (Table 5). For unrotated functions *f1* to *f8*, PPSO-*gbest*-SE performs better than the rest of the algorithms. For the rotated benchmark functions, results of PPSO algorithms are quite comparable with those by the other PSO techniques except for EPUS-PSO which has the best performance.

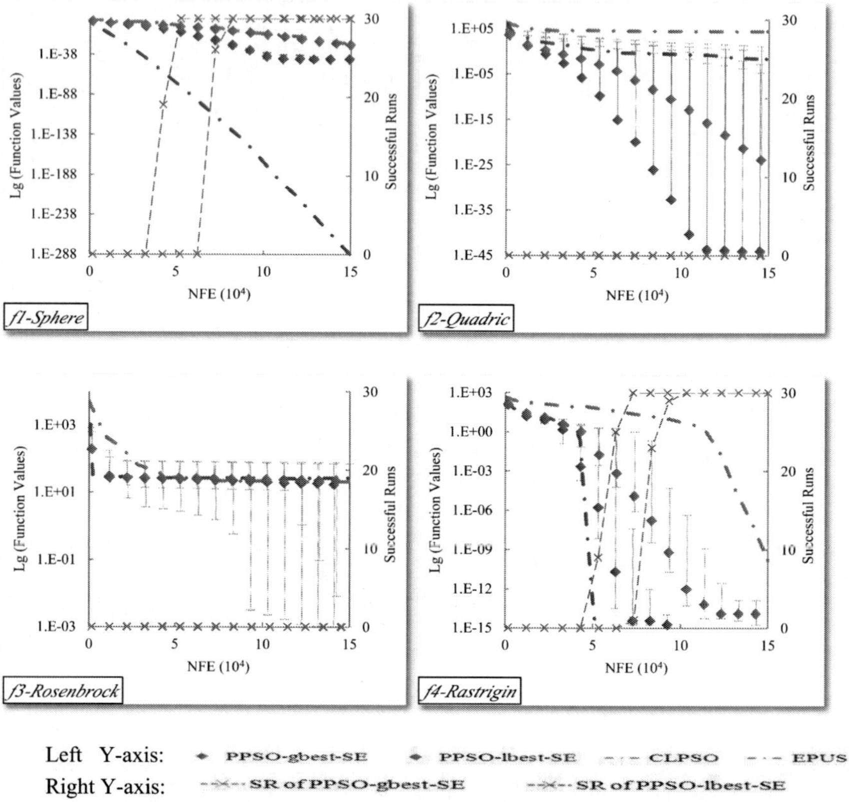

Figure 5. Median convergence profiles for 30 dimensional benchmark functions.

The median convergence profile is plotted for all the benchmark functions using PPSO-*gbest*-SE, PPSO-*lbest*-SE, EPUS-PSO and CLPSO algorithms in Figures 5 and 6 for 30-D and 10-D problems respectively. The profiles for EPUS-PSO and CLPSO are extracted from Hsieh et al. [17]. In these figures, error bars and number of successful runs are also included for PPSO-*gbest*-SE and PPSO-*lbest*-SE. The upper and lower error bars represent respectively the maximum and minimum objective function values at different NFE levels from 30 different runs. Error bars give an indication of the consistency in the runs. The larger the error bars, the more inconsistent the runs are, which results in higher standard deviation for the mean of the objective function values obtained. Taking 10-D *f12-R.Non-Continuous Rastrigin* as an example, its errors bars are large, which explains the high standard deviation for the final objective function value obtained as shown in Table 4.

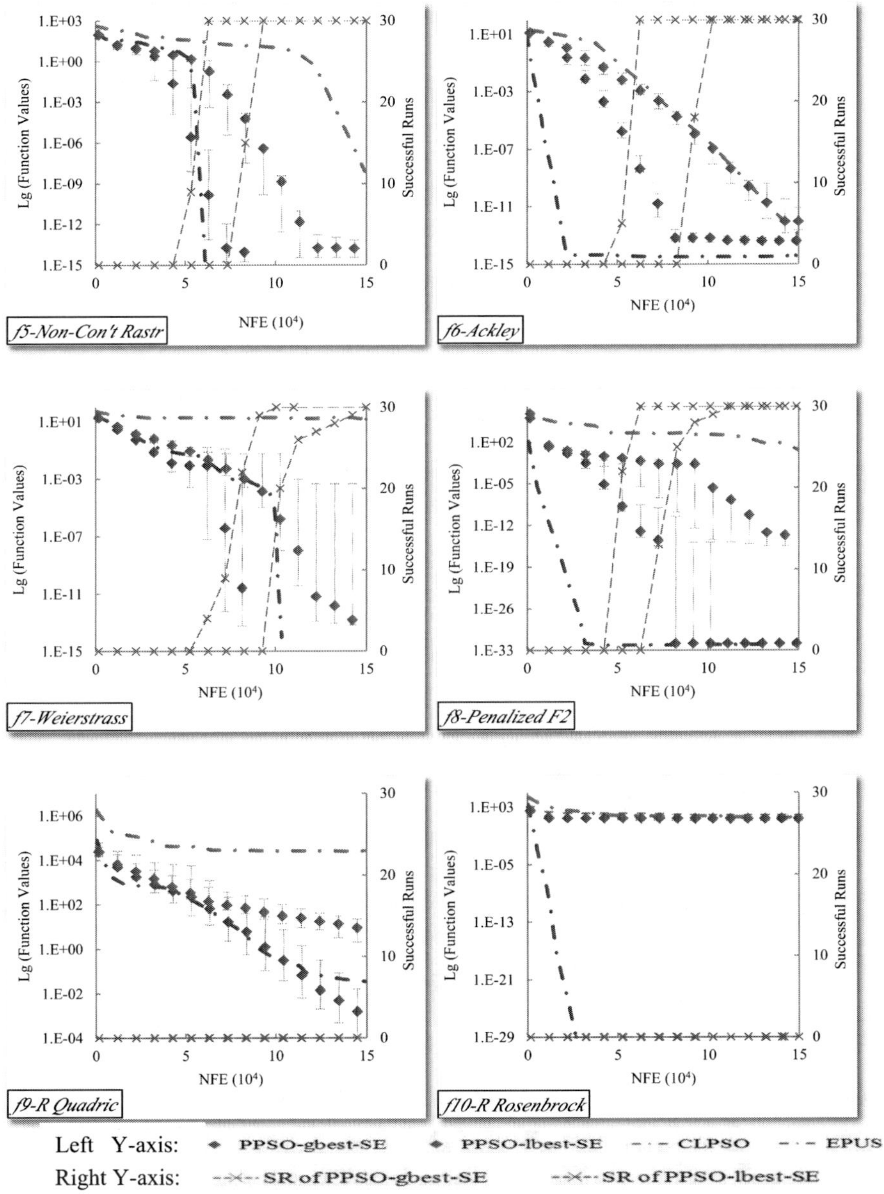

Figure 5. (continued). Median Convergence Profiles for 30-D Benchmark Functions.

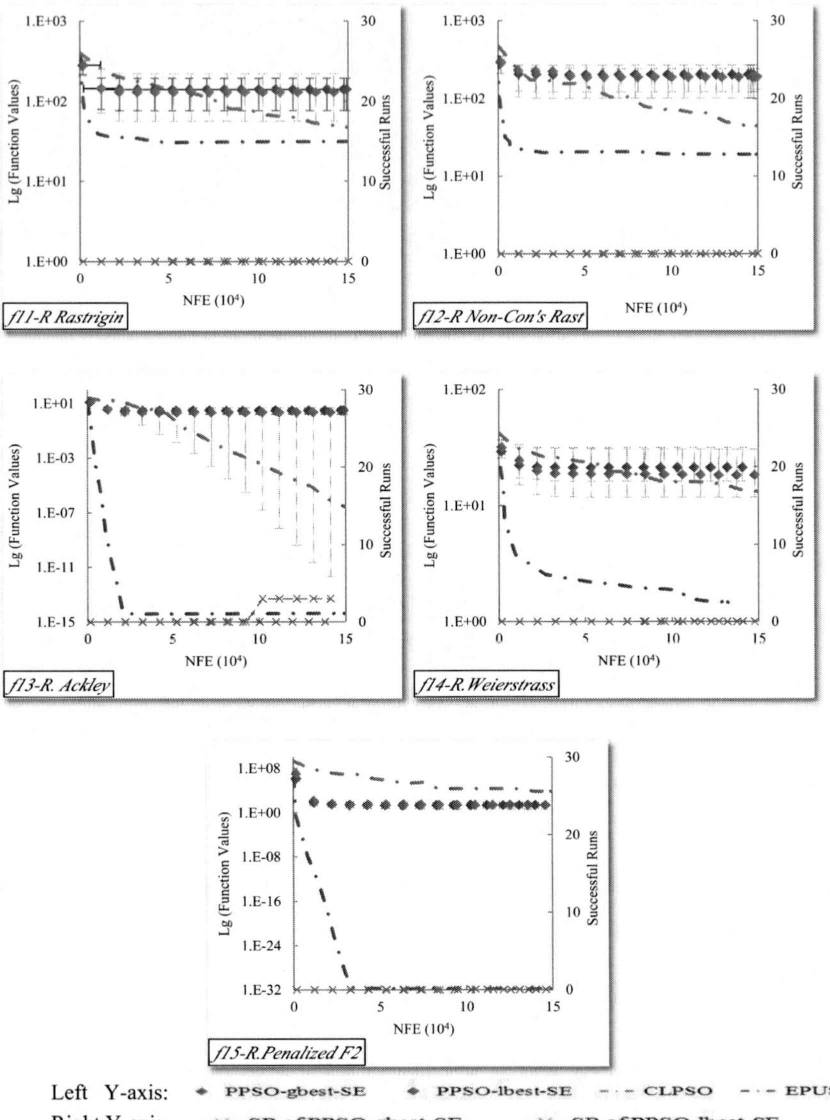

Figure 5. Median Convergence Profiles for 30-D Benchmark Functions.

Table 4. Results of Different PSO Techniques on 10-D Benchmark Functions

Benchmark Function	EPUS-PSO	CLPSO	CPSO	UPSO	PPSO-*gbest*-SE		PPSO-*lbest*-SE		UBBPSO	
	Mean (Std)	Mean (Std)	Mean (Std)	Mean (Std)	Mean (Std)	SR (%)	Mean (Std)	SR (%)	Mean (Std)	SR (%)
f1- Sphere.	5.55E-153 (2.44E-134)	4.02E-21 (6.49E-21)	9.01E-10 (1.38E-09)	5.52E-90 (1.18E-89)	2.36E-45 (7.01E-46)	100	7.86E-34 (3.16E-33)	100	4.97E-45 (1.38E-45)	100
f2- Quadric.	9.13E-12 (4.16E-11)	5.04E+02 (1.99E+02)	1.62E+03 (9.35E+02)	7.18E+01 (7.63E+01)	1.27E-21 (6.65E-21)	100	4.80E-05 (6.71E-05)	13	8.89E-09 (3.40E-08)	100
f3- Rosenbrock.	9.13E-02 (1.05E-01)	2.84E+00 (1.05E+00)	1.08E+00 (1.18E+00)	2.89E-01 (7.36E-02)	1.04E+00 (6.13E-01)	0	3.76E+00 (1.04E+00)	0	4.45E+00 (1.00E+00)	0
f4- Rastrigin.	0.00E+00 (0.00E+00)	2.35E-09 (2.72E-09)	3.83E-10 (7.36E-10)	9.28E+00 (2.97E+00)	0.00E+00 (0.00E+00)	100	4.03E-15 (9.31E-15)	100	3.30E-11 (1.80E-11)	100
f5- Non- Continuous Rastrigin.	0.00E+00 (0.00E+00)	6.58E-09 (1.75E-08)	4.09E-10 (1.18E-09)	5.64E+00 (1.58E+00)	0.00E+00 (0.00E+00)	100	4.03E-15 (6.15E-15)	100	3.00E-12 (1.83E-11)	100
f6- Ackley.	2.72E-15 (1.50E-15)	1.78E-11 (1.63E-11)	7.02E-06 (5.77E-06)	3.32E-15 (8.86E-16)	1.11E-14 (4.57E-15)	100	9.65E-14 (9.93E-14)	100	3.73E-15 (1.01E-15)	100
f7- Weierstrass.	3.21E-02 (8.11E-02)	2.06E+00 (4.80E-01)	4.12E+00 (1.72E+00)	1.94E+00 (1.13E+00)	0.00E+00 (0.00E+00)	100	9.36E-15 (2.35E-14)	100	0.00E+00 (0.00E+00)	100
f8- Penalized Fun. 2.	1.35E-32 (5.47E-48)	4.32E-01 (4.10E-01)	1.09E+06 (2.41E+06)	1.35E-32 (5.47E-48)	1.35E-32 (0.00E+00)	100	1.89E-16 (5.95E-16)	100	3.66E-14 (2.01E-13)	100
f9- R.Quadric.	3.99E-07 (1.63E-06)	8.27E+02 (3.35E+02)	2.15E+03 (1.19E+03)	1.01E+02 (9.50E+01)	7.93E-20 (4.33E-19)	100	7.36E-05 (1.42E-04)	7	1.61E-08 (3.20E-08)	100
f10- R.Rosenbrock.	8.96E-01 (2.15E+00)	6.44E+00 (4.56E+00)	3.01E+00 (2.52E+00)	1.11E+00 (1.18E+00)	2.94E+00 (1.76E+00)	0	5.88E+00 (1.74E+00)	0	5.74E+00 (1.42E+00)	0
f11- R.Rastrigin.	7.26E+00 (4.31E+00)	8.85E+00 (3.07E+00)	2.81E+01 (1.39E+01)	1.18E+01 (5.66E+00)	2.32E+01 (9.35E+00)	0	1.83E+01 (8.80E+00)	0	1.57E+01 (3.95E+00)	0
f12- R.Non-Continuous Rastrigin.	6.40E+00 (2.41E+00)	7.07E+00 (2.02E+00)	1.96E+01 (1.08E+01)	1.05E+01 (3.63E+00)	3.37E+01 (1.43E+01)	0	2.86E+01 (9.37E+00)	0	2.11E+01 (4.48E+00)	0

Table 4. (Continued)

Benchmark Function	EPUS-PSO	CLPSO	CPSO	UPSO	PPSO-*gbest*-SE		PPSO-*lbest*-SE		UBBPSO	
	Mean (Std)	Mean (Std)	Mean (Std)	Mean (Std)	Mean (Std)	SR (%)	Mean (Std)	SR (%)	Mean (Std)	SR (%)
f13- R.Ackley.	2.61E-15 (1.57E-15)	8.45E-08 (2.97E-07)	1.31E+01 (9.68E-01)	7.70E-02 (2.88E-01)	7.55E-01 (9.91E-01)	60	9.34E-02 (3.61E-01)	93	3.65E-15 (1.38E-15)	100
f14- R.Weiertrass.	1.12E+00 (1.36E+00)	1.98E+00 (5.38E-01)	3.48E+00 (1.41E+00)	1.84E+00 (1.02E+00)	2.66E+00 (1.66E+00)	0	2.77E+00 (1.80E+00)	0	6.92E-01 (9.86E-01)	0
f15- R.Penalized.Fun.2.	1.35E-32 (5.47E-48)	4.36E-01 (5.55E-01)	1.09E+06 (2.37E+06)	2.77E+00 (6.22E+00)	8.32E-03 (2.47E-02)	80	7.33E-04 (2.79E-03)	93	8.84E-15 (4.23E-14)	100

Table 5. Results of Different PSO Techniques on 30-D Benchmark Functions

Benchmark Function	EPUS-PSO	CLPSO	CPSO	UPSO	PPSO-*gbest*-SE		PPSO-*lbest*-SE		UBBPSO	
	Mean (Std)	Mean (Std)	Mean (Std)	Mean (Std)	Mean (Std)	SR (%)	Mean (Std)	SR (%)	Mean (Std)	SR (%)
f1- Sphere.	8.50E-263 (1.02E-262)	1.34E-25 (1.71E-25)	1.57E-10 (2.99E-10)	8.79E-123 (3.56E-122)	9.04E-45 (5.98E-45)	100	9.48E-26 (2.36E-25)	100	5.71E-44 (5.67E-44)	100
f2- Quadric.	1.97E+00 (4.69E+00)	2.41E+04 (6.78E+03)	8.36E+04 (6.58E+04)	1.07E+04 (4.23E+03)	1.12E-04 (1.28E-04)	0	7.03E+00 (3.00E+00)	0	2.49E+01 (1.25E+01)	0
f3- Rosenbrock.	2.55E+01 (3.53E-01)	2.01E+01 (3.37E+00)	1.03E+01 (1.37E+01)	1.31E+01 (2.98E+00)	2.44E+01 (2.51E+01)	0	2.87E+01 (2.15E+01)	0	2.18E+01 (8.76E-01)	0
f4- Rastrigin.	0.00E+00 (0.00E+00)	1.87E-09 (5.34E-09)	3.15E-10 (1.05E-09)	7.63E+01 (1.45E+01)	0.00E+00 (0.00E+00)	100	1.22E-14 (1.49E-14)	100	6.96E-10 (2.68E-10)	100
f5- Non- Continuous Rastrigin.	0.00E+00 (0.00E+00)	2.02E-07 (2.35E-07)	8.84E-11 (2.21E-10)	8.40E+01 (1.62E+01)	0.00E+00 (0.00E+00)	100	1.67E-14 (1.52E-14)	100	0.00E+00 (0.00E+00)	100
f6- Ackley.	3.91E-15 (1.07E-15)	9.90E-14 (3.80E-14)	3.90E-06 (5.98E-06)	1.33E+00 (8.58E-01)	4.21E-14 (8.45E-15)	100	1.29E-12 (1.82E-12)	100	5.18E-14 (1.70E-14)	100

Table 5. (Continued)

Benchmark Function	EPUS-PSO	CLPSO	CPSO	UPSO	PPSO-*gbest*-SE		PPSO-*lbest*-SE		UBBPSO	
	Mean (Std)	Mean (Std)	Mean (Std)	Mean (Std)	Mean (Std)	SR (%)	Mean (Std)	SR (%)	Mean (Std)	SR (%)
f7- *Weierstrass.*	4.08E-01 (1.22E+00)	1.49E+01 (1.69E+00)	1.36E+01 (3.38E+00)	2.06E+01 (3.37E+00)	0.00E+00 (0.00E+00)	100	3.93E-06 (2.15E-05)	97	0.00E+00 (0.00E+00)	100
f8- *Penalized Fun. 2.*	1.35E-32 (5.47E-48)	1.62E+03 (5.92E+03)	2.58E+08 (4.41E+08)	1.34E+01 (2.34E+01)	1.35E-32 (0.00E+00)	100	5.09E-15 (4.13E-15)	100	1.35E-32 (0.00E+00)	100
f9- *R.Quadric.*	1.20E+01 (2.46E+01)	2.38E+04 (6.49E+03)	3.87E+04 (1.71E+04)	1.27E+04 (4.77E+03)	1.21E-03 (1.71E-03)	7	7.98E+00 (3.54E+00)	0	5.68E-04 (1.33E-04)	97
f10- *R.Rosenbrock.*	1.96E+01 (2.91E+01)	3.04E+01 (1.32E+01)	3.03E+01 (2.59E+01)	2.15E+01 (1.89E+01)	2.70E+01 (1.42E+01)	0	3.09E+01 (1.74E+01)	0	2.36E+01 (1.94E+00)	0
f11- *R.Rastrigin.*	3.76E+01 (1.48E+01)	4.70E+01 (6.85E+00)	9.52E+01 (2.93E+01)	7.70E+01 (1.70E+01)	1.38E+02 (2.77E+01)	0	1.29E+02 (3.11E+01)	0	1.81E+02 (1.46E+01)	0
f12- *R.Non-Continuous Rastrigin.*	3.03E+01 (1.11E+00)	4.05E+01 (7.91E+00)	8.38E+01 (2.68E+01)	9.09E+01 (1.94E+01)	2.04E+02 (3.55E+01)	0	1.88E+02 (2.95E+01)	0	1.79E+02 (1.96E+01)	0
f13- *R.Ackley.*	6.81E-01 (1.02E+00)	2.22E-06 (8.29E-06)	1.52E+00 (3.80E-01)	1.63E+00 (9.49E-01)	3.05E+00 (7.66E-01)	0	2.26E+00 (1.06E+00)	3	4.48E-15 (1.64E-15)	100
f14- *R.Weiertrass.*	4.26E+00 (2.97E+00)	1.51E+01 (1.93E+00)	1.36E+01 (3.56E+00)	2.02E+01 (4.19E+00)	2.24E+01 (4.68E+00)	0	2.00E+01 (3.81E+00)	0	7.16E+00 (2.46E+00)	0
f15- *R.Penalized Fun. 2.*	1.35E-32 (5.47E-48)	3.43E+04 (1.57E+05)	2.44E+08 (4.66E+08)	2.74E+01 (2.77E+01)	2.17E+01 (1.50E+01)	0	2.33E+01 (1.51E+01)	0	9.02E-01 (3.62E+00)	0

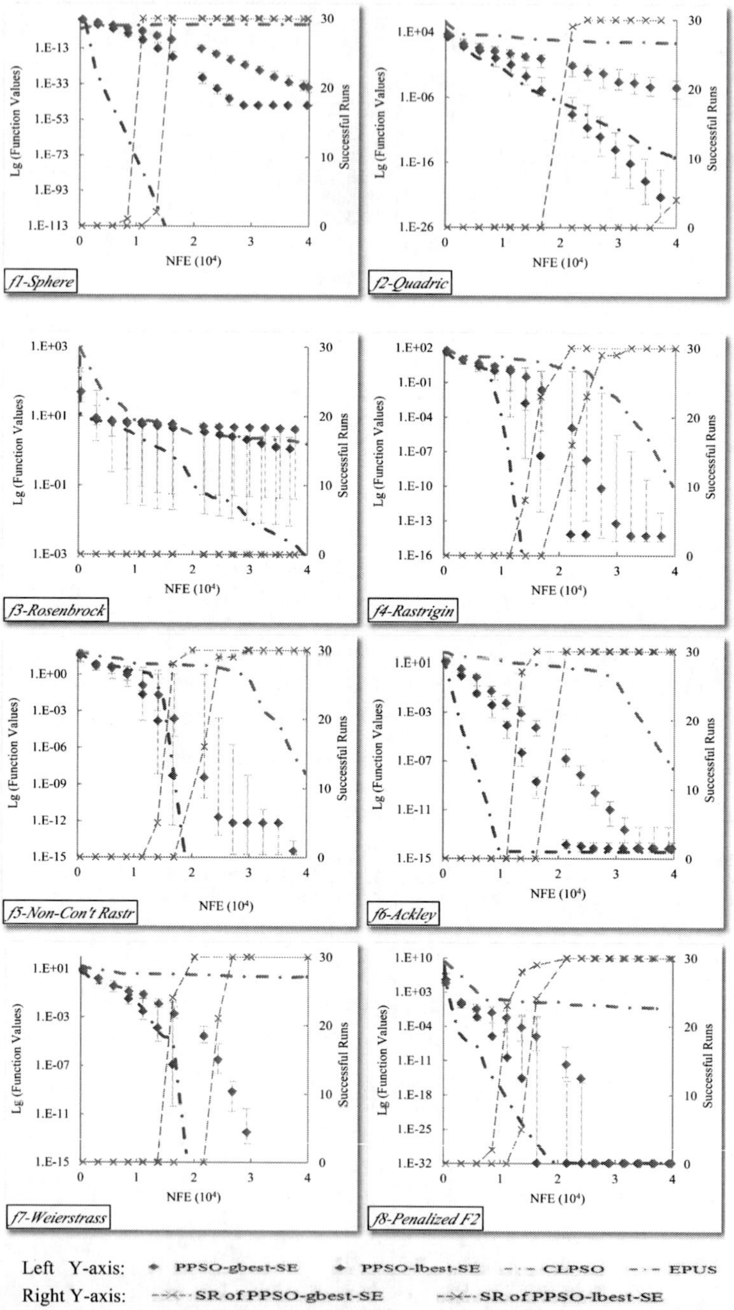

Figure 6. (Continued).

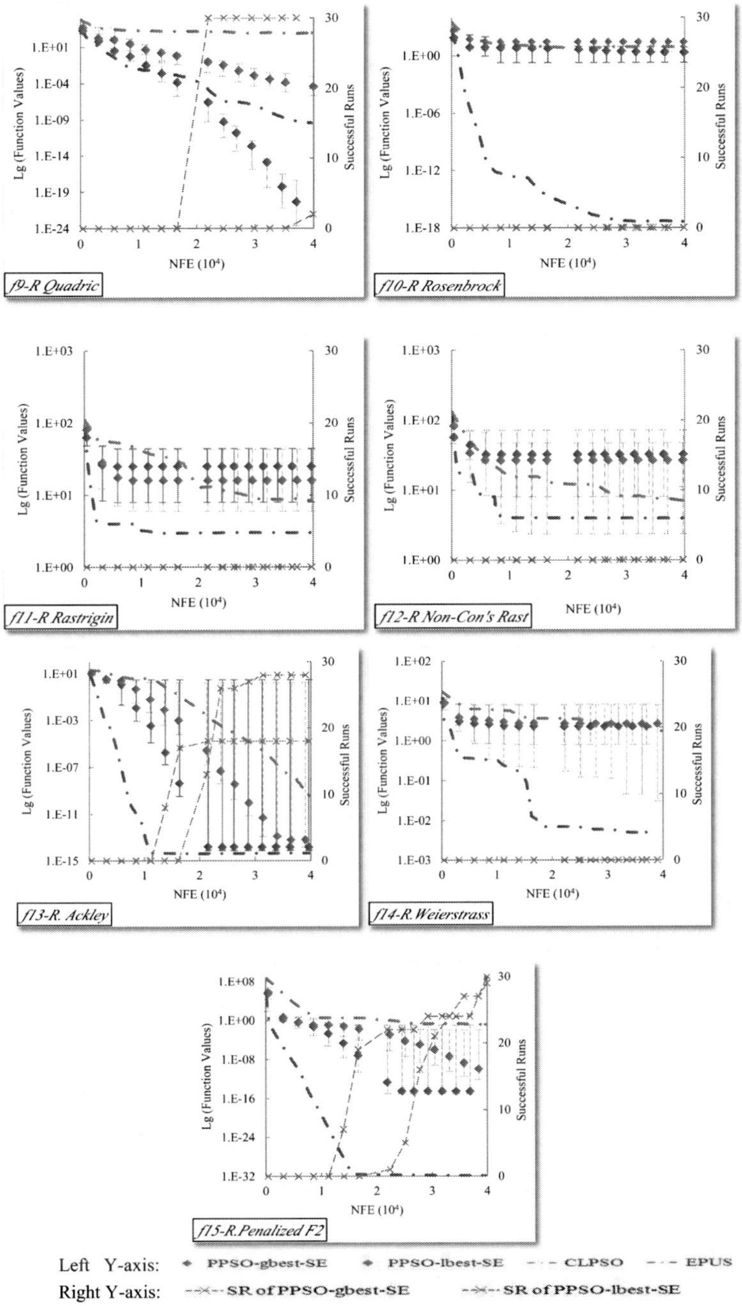

Figure 6. Median Convergence Profiles for 10-D Benchmark Functions.

From the convergence profiles, PPSO-*gbest*-SE generally converges faster than PPSO-*lbest*-SE. This is consistent with the better results given in Table 4 for PPSO-*gbest*-SE. So, PPSO-*gbest*-SE requires lower NFE to achieve the same number of successful runs as compared to PPSO-*lbest*-SE (Figures 5 and 6).

Both algorithms have comparable error bars for all benchmark functions tested. Comparing the convergence profiles of EPUS-PSO and PPSO-*gbest*-SE, it is observed that EPUS-PSO shows better convergence profile for most of the benchmark functions, especially for rotated functions. On the other hand, PPSO-*gbest*-SE and PPSO-*lbest*-SE show better convergence profiles than CLPSO. These observations are consistent with the results for the mean value of the run presented in Table 4.

4. 4. Comparison of PPSO with UBBPSO Algorithm

Compared to PSO, bare-bones PSO (BBPSO) has no tuning parameter other than the population size [30]. Recently, Zhang et al. [31] proposed a new BBPSO with crossover and mutation operators of differential evolution, and shown it to be better than or comparable to other stochastic global optimization algorithms tested for parameter estimation in modeling vapor-liquid equilibrium data.

Then, Zhang et al. [12] combined the global and local topologies in this new BBPSO to develop the unified BBPSO (UBBPSO) and tested it for phase equilibrium and stability problems. Flowchart of UBBPSO is available in Zhang et al. [12].

Results of solving the benchmark functions *f1* to *f15* by UBBPSO are also shown in Table 4 (for 10-D problems) and Table 5 (for 30-D problems). For functions *f1* to *f12*, PPSO-*gbest*-SE has similar performance compared to UBBPSO. However, UBBPSO performs better than PPSO-*gbest*-SE for function *f13* to *f15*.

Conclusion

In this study, two re-initialization strategies: *selection* and *escape* strategies have been introduced to PSO algorithm with the aim of discouraging the population from being trapped in the local minima. *Selection* strategy works on the individual particle level while *escape* strategy works on the population level. The former strategy is a "greedy" strategy while the latter is not. Re-initialization for both strategies is done by assigning the new particle to *gbest* position and randomly changing one of the dimensions of the new particle.

From the results obtained, PPSO-E outperforms PPSO-S. As both re-initialized strategies have unique strengths, a hybrid approach of both strategies has been proposed, PPSO-*gbest*-SE. The performance of this hybrid algorithm is relatively better than the algorithm with individual re-initialization strategy.

The effect of topology on this hybrid approach has been studied, and fully connected network, PPSO-*gbest*-SE is found to be better. Based on the performance results for 10-D and 30-D benchmark functions, PPSO-*gbest*-SE is better than CLPSO, CPSO and UPSO; however, EPUS-PSO and UBBPSO seem to be slightly better than PPSO-*gbest*-SE, particularly for rotated functions.

APPENDIX A: ORTHOGONAL MATRIX

Table A1. Orthogonal Matrix (M) for 10-D, Conditioned Number of One

	1	2	3	4	5	6	7	8	9	10
1	0.18	-0.37	0.09	0.64	0.05	0.14	0.13	0.29	-0.32	0.43
2	-0.07	0.00	0.12	0.03	-0.69	-0.05	0.00	0.59	0.39	-0.05
3	0.03	-0.22	-0.34	0.46	0.11	-0.61	-0.18	-0.04	0.21	-0.40
4	-0.56	-0.26	-0.31	-0.13	0.18	0.13	0.54	0.28	-0.13	-0.26
5	0.09	-0.31	-0.11	0.02	-0.34	0.46	-0.38	-0.13	-0.37	-0.50
6	-0.22	0.01	0.74	0.10	-0.11	-0.34	0.19	-0.14	-0.35	-0.30
7	0.70	0.13	0.10	-0.05	0.21	0.11	0.41	0.24	0.06	-0.44
8	0.25	-0.71	0.07	-0.54	-0.04	-0.31	0.01	-0.01	0.00	0.18
9	-0.01	0.23	-0.05	-0.22	0.25	-0.24	-0.45	0.60	-0.46	-0.02
10	0.19	0.26	-0.44	-0.05	-0.49	-0.31	0.34	-0.16	-0.46	0.13

Table A2. Orthogonal Matrix (M) for 30-D, Conditioned Number of One

	1	2	3	4	5	6	7	8	9	10	11	12	13	14	15
1	-0.02	0.31	-0.14	0.02	0.16	-0.12	0.18	-0.29	-0.17	0.03	0.01	0.13	0.01	0.15	0.25
	0.19	0.21	-0.06	0.27	-0.09	0.13	0.34	-0.12	0.23	-0.17	0.04	0.21	0.00	-0.04	0.23
3	0.09	0.29	-0.13	-0.08	-0.22	-0.10	0.12	0.23	-0.09	-0.17	-0.04	0.11	-0.27	0.21	-0.28
4	0.15	0.30	-0.15	-0.20	0.15	-0.29	0.11	-0.26	0.17	0.15	-0.15	-0.15	-0.29	-0.21	-0.16
5	-0.24	-0.03	-0.14	0.20	0.08	0.04	0.21	0.04	-0.08	0.02	0.10	-0.15	-0.37	0.04	0.05
6	0.14	0.04	0.20	-0.56	0.18	-0.12	0.01	0.20	-0.28	-0.16	-0.03	0.21	-0.08	0.33	0.02
7	0.43	0.08	0.13	0.06	0.07	-0.03	0.18	0.03	0.06	-0.14	-0.08	-0.22	0.38	-0.19	-0.19
8	-0.01	-0.07	-0.03	-0.13	-0.06	-0.07	0.27	-0.32	-0.09	-0.19	0.02	-0.45	-0.03	0.12	0.06
9	-0.04	0.28	0.21	-0.07	0.39	0.07	-0.09	-0.03	0.40	-0.16	0.20	-0.17	0.18	0.16	0.08
10	0.08	-0.04	-0.10	-0.09	0.16	-0.09	-0.13	-0.12	-0.16	-0.07	0.31	0.19	-0.09	-0.32	0.03
11	-0.12	0.48	-0.03	0.06	-0.07	0.20	-0.11	-0.08	-0.39	0.25	-0.16	0.11	0.16	-0.22	-0.02
12	0.18	0.00	0.18	-0.14	-0.20	0.10	-0.09	-0.34	0.06	0.01	-0.02	-0.07	-0.19	0.03	-0.02
13	0.24	-0.13	-0.20	0.08	-0.06	-0.28	-0.07	-0.08	0.01	0.10	0.41	0.07	0.24	0.37	0.12
14	0.00	-0.17	-0.35	0.15	0.30	-0.30	-0.31	-0.18	-0.05	0.02	-0.16	0.02	-0.05	0.03	0.04
15	-0.01	0.08	0.02	-0.01	0.16	-0.18	0.00	0.03	-0.15	0.04	0.32	0.08	0.25	-0.33	-0.13
16	-0.25	0.02	-0.31	0.15	-0.13	-0.10	-0.10	0.06	0.09	-0.47	-0.10	-0.07	0.20	0.03	-0.25
17	0.12	-0.02	0.07	-0.05	0.17	0.20	-0.13	0.18	0.02	0.35	-0.13	-0.22	0.12	0.08	0.17
18	0.27	0.24	-0.01	0.05	-0.19	-0.02	-0.17	0.04	-0.04	-0.14	0.34	-0.06	-0.14	-0.15	0.21
19	0.03	0.18	0.24	0.31	0.30	-0.08	-0.04	-0.07	-0.07	-0.10	-0.18	0.22	-0.03	0.27	-0.14
20	-0.25	-0.04	-0.21	-0.23	0.28	0.39	0.04	-0.02	-0.12	-0.35	0.12	0.04	0.04	-0.09	0.22
21	-0.17	-0.22	0.33	0.02	0.06	-0.20	0.16	-0.29	0.14	-0.06	-0.21	0.40	0.08	-0.13	0.07
22	-0.28	0.18	-0.01	-0.08	-0.10	0.25	0.10	-0.28	0.11	0.19	0.19	0.03	0.13	0.26	-0.11
23	-0.36	0.17	0.26	-0.17	-0.23	-0.25	-0.41	-0.14	0.17	-0.05	0.14	-0.09	-0.02	-0.07	-0.05

Table A2. (Continued)

	1	2	3	4	5	6	7	8	9	10	11	12	13	14	15
24	-0.06	0.04	0.16	0.33	-0.06	0.01	-0.01	0.09	-0.03	0.14	0.30	0.12	-0.12	0.19	-0.10
25	0.14	-0.03	0.04	0.09	0.07	0.16	-0.27	0.04	0.27	-0.08	-0.05	0.09	-0.38	-0.10	0.26
26	-0.20	-0.09	0.01	-0.18	0.00	-0.23	0.42	0.24	0.21	0.25	0.18	0.10	-0.05	-0.19	0.05
27	-0.13	0.18	0.18	0.15	0.22	-0.09	0.06	0.31	0.06	-0.15	0.06	-0.09	-0.15	-0.11	-0.07
28	-0.15	0.09	0.00	0.14	0.21	-0.21	0.01	0.08	-0.08	0.19	-0.01	-0.35	-0.04	0.09	0.13
29	0.11	-0.06	-0.25	-0.10	0.27	0.28	-0.05	-0.09	0.25	0.17	0.15	0.14	-0.10	0.04	-0.55
30	0.05	-0.25	0.34	0.18	0.12	0.11	0.08	-0.23	-0.37	-0.10	0.23	-0.18	-0.18	-0.07	-0.26
	16	17	18	19	20	21	22	23	24	25	26	27	28	29	30
1	0.37	-0.02	-0.11	0.18	0.00	0.03	-0.17	-0.22	-0.15	0.02	0.11	0.17	-0.06	0.42	0.28
2	-0.30	-0.03	0.20	-0.10	0.10	-0.13	0.31	-0.04	-0.13	0.17	-0.31	0.06	0.29	0.03	0.10
3	0.29	-0.30	0.44	-0.01	0.13	-0.20	0.05	-0.05	0.20	-0.11	0.08	0.03	-0.11	-0.16	-0.01
4	-0.04	0.33	0.12	0.12	-0.30	0.06	-0.05	-0.04	0.07	0.19	0.02	-0.12	0.14	-0.09	-0.26
5	-0.15	0.06	-0.20	-0.23	-0.13	-0.08	0.39	-0.19	0.01	-0.26	0.46	-0.21	0.01	0.02	0.07
6	-0.11	0.12	-0.25	-0.10	-0.04	-0.10	0.26	0.11	-0.03	0.02	-0.16	-0.06	0.19	0.15	-0.02
7	-0.03	0.05	-0.10	0.07	-0.19	-0.18	0.09	-0.09	0.27	-0.43	-0.06	-0.03	-0.19	0.20	0.13
8	0.15	-0.12	0.01	-0.29	0.18	0.43	0.01	0.16	0.00	-0.09	-0.30	-0.22	-0.07	-0.09	0.07
9	0.20	0.05	-0.01	0.06	0.11	-0.15	0.10	0.22	-0.10	0.07	0.26	-0.04	0.00	-0.37	0.16
10	0.06	-0.01	0.09	-0.46	-0.35	-0.12	-0.09	0.07	-0.17	-0.14	-0.12	0.31	-0.16	-0.24	0.16
11	-0.04	-0.01	-0.05	-0.06	0.04	-0.09	0.13	0.29	-0.01	0.09	-0.06	-0.41	-0.21	-0.07	0.11
12	-0.28	-0.29	0.08	0.07	-0.12	0.05	0.09	0.44	0.02	0.05	0.36	0.26	-0.13	0.28	0.06
13	-0.17	0.06	0.18	-0.09	-0.09	-0.09	-0.06	-0.02	0.10	0.22	0.11	-0.38	-0.28	0.08	-0.07
14	-0.08	-0.02	0.14	0.15	0.28	-0.18	0.06	0.31	0.05	-0.38	-0.09	-0.03	0.26	0.05	0.09
15	0.08	-0.04	0.09	-0.11	0.33	0.32	0.34	-0.02	0.21	0.11	0.22	0.19	0.13	0.16	-0.26

Table A2. (Continued)

	1	2	3	4	5	6	7	8	9	10	11	12	13	14	15
16	0.17	0.00	-0.10	-0.04	-0.30	-0.04	0.17	0.22	-0.29	0.17	-0.02	0.01	-0.01	0.25	-0.20
17	0.21	-0.20	0.44	-0.18	-0.26	0.06	0.13	-0.07	-0.29	-0.11	0.00	-0.11	0.21	0.22	-0.14
18	0.02	-0.32	-0.32	0.19	0.00	0.01	-0.18	-0.02	-0.19	-0.23	0.00	-0.20	0.22	-0.06	-0.34
19	-0.27	-0.03	0.06	0.02	-0.01	0.35	-0.01	-0.11	-0.24	-0.21	-0.03	0.07	-0.32	-0.13	-0.29
20	-0.19	-0.10	0.27	0.33	-0.20	0.10	-0.05	-0.05	0.28	-0.04	-0.04	-0.10	-0.09	0.01	-0.10
21	0.21	-0.36	-0.05	-0.11	-0.17	-0.13	-0.02	-0.01	0.19	0.02	0.08	-0.31	0.13	-0.04	-0.13
22	0.00	0.18	-0.04	-0.23	0.06	-0.26	-0.15	0.07	0.14	-0.31	-0.11	0.17	0.08	0.12	-0.42
23	-0.17	-0.01	0.17	0.06	-0.01	0.00	0.16	-0.38	-0.04	-0.12	-0.20	-0.10	-0.03	0.16	0.23
24	0.20	0.11	0.00	0.21	-0.39	0.30	0.11	0.29	0.29	-0.07	-0.21	0.00	0.23	-0.02	0.20
25	0.31	0.19	-0.06	-0.13	0.16	-0.01	0.18	0.04	0.21	0.05	-0.19	-0.07	-0.39	0.26	-0.18
26	-0.02	-0.04	0.06	0.27	0.07	-0.03	0.08	0.28	-0.32	-0.20	-0.13	-0.05	-0.32	0.09	-0.05
27	-0.21	-0.02	0.10	-0.30	0.08	-0.01	-0.53	0.18	0.09	0.09	0.04	-0.14	0.12	0.36	0.12
28	-0.11	-0.42	-0.20	0.03	-0.11	-0.25	0.11	-0.04	0.24	0.28	-0.29	0.28	-0.13	-0.06	-0.13
29	-0.03	-0.33	-0.22	-0.02	0.06	0.08	0.02	-0.15	-0.07	0.10	-0.16	-0.16	0.00	0.11	0.17
30	0.13	0.13	0.18	0.22	0.11	-0.36	0.03	-0.04	-0.23	0.17	-0.08	-0.10	0.01	0.08	-0.10

REFERENCES

[1] C. Blum, A. Roli, *Metaheuristics in combinatorial optimization: overview and conceptual comparison, A. C. M. Comput. Surv.*, 35, 2003, 268-308.

[2] G. P. *Rangaiah (Ed.), Stochastic Global Optimization: Techniques and applications in Chemical Engineering, World Scientific*, 2010.

[3] R. C. Eberhart, J. Kennedy, *A New Optimizer Using Particle Swarm Theory, Proceedings of the Sixth International Symposium, Micromachine Human Science*, 1995, 39–43.

[4] R. C. Eberhart, J. Kennedy, *Particle Swarm Optimization, Proceedings of the I. E. E. E. International Conference Neural Networks*, 1995, 1942–1948.

[5] K. E. Parsopoulos, M. N. Vrahatis, *Recent approaches to global optimization problems through particle swarm optimization, Natural Computing,* 1, 2002, 235-306.

[6] M. M. Ali, P. Kaelo, *Improved particle swarm optimization algorithms for global optimization, Applied Mathematics and Computation,* 196, 2008, 578-593.

[7] A. Nickabadi, M. M. Ebadzadeh, R. Safabakhsh, *A novel particle swarm optimization algorithm with adaptive inertia weight, Applied Soft Computing,* 11, 2011, 3658-3670.

[8] H. Gao, W. Xu, *Particle swarm algorithm with hybrid mutation strategy, Applied Soft Computing,* 11, 2011, 5129-5142.

[9] A. Bonilla-Petriciolet, J. G. Segovia-Hernandez, *A comparative study of particle swarm optimization and its variants for phase stability and equilibrium calculations in multicomponente reactive and non-reactive systems, Fluid Phase Equilibria,* 289, 2010, 110-121.

[10] A. Bonilla-Petriciolet, G. P. Rangaiah, J. G. Segovia-Hernandez, *Evaluation of stochastic global optimization methods for modeling vapor-liquid equilibrium data, Fluid Phase Equilibria,* 287, 2010, 111-125.

[11] R. Thangaraj, M. Pant, A. Abraham, P. Bouvry, *Particle swarm optimization: hybridization perspectives and experimental illustrations, Applied Mathematics and Computation,* 217, 2011, 5208-5226.

[12] H. Zhang, J. A. Fernandez-Vargas, G. P. Rangaiah, A. Bonilla-Petriciolet, J. G. Segovia-Hernandez, *Evaluation of integrated differential evolution and unified bare-bones particle swarm*

optimization for phase equilibrium and stability problems, *Fluid Phase Equilibria*, 310, 2011, 129-141.

[13] T. Y. Chen, T. M. Chi, *On the improvements of the particle swarm optimization algorithm, Advances in Engineering Software,* 41, 2010, 229-239.

[14] H. Liu, Z. Cai, Y. Wang, *Hybridizing particle swarm optimization with differential evolution for constrained numerical and engineering optimization, Applied Soft Computing,* 10, 2010, 629-640.

[15] Y. Kao, E. Zahara, *A hybrid genetic algorithm and particle swarm optimization for multimodal functions*, *Applied Soft Computing,* 8, 2008, 849-857,

[16] M. R. Chen, X. Li, X. Zhang, Y. Z. Lu, *A novel particle swarm optimizer hybridized with extremal optimization, Applied Soft Computing,* 10, 2010, 367-373.

[17] S. T. Hsieh, T. Y. Sun, C. C. Liu, S. J. Tsai, *Efficient Population Utilization Strategy for Particle Swarm Optimizer, I. E. E. E. Transactions Cybernetics,* 39, 2009, 444-456.

[18] J. J. Liang, A. K. Qin, P. N. Suganthan, S. Baskar, *Comprehensive Learning Particle Swarm Optimizer for Global Optimization of Multimodal Functions*, *IEEE Transactions on Evolutionary Computation,* 10, 2006, 281–296.

[19] F. van den Bergh, A. P. Engelbrecht, *A Cooperative Approach to Particle Swarm Optimization*, *IEEE Transactions on Evolutionary Computation,* 8, 2004, 225–239.

[20] K. E. Parsopoulos, M. N. Vrahatis, *Unified Particle Swarm Optimization in Dynamic Environments*. *Lecture Notes in Computer Science,* 3449, 2005, 590–599.

[21] R. C. Eberhart, Y. Shi, *A Modified Particle Swarm Optimizer, IEEE World Congress on Computational Intelligence*, 1998, 69-73.

[22] R. C. Eberhart, Y. Shi, *Comparing Inertia Weights and Constriction Factors in Particle Swarm Optimization*, *Proceedings of the IEEE Congress on Evolutionary Computation*, 2000, 84–88.

[23] R. C. Eberhart, Y. Shi, *Tracking and Optimizing Dynamic Systems with Particle Swarms*, *Proceedings of the IEEE Congress on Evolutionary Computation*, 2001, 94–100.

[24] Y. L. Zheng, L. H. Ma, L. Y. Zhang, J. X.Qian, *On the Convergence Analysis and Parameter Selection in Particle Swarm Optimization*, *Proceedings of the I. E. E. E. International Conference on Machine Learning and Cybernetics*, 2003, 1802–1807.

[25] J. Kennedy, *The Behavior of Particles, Proceedings of the Seventh Annual Conference on Evolutionary Programming*, 1998, 581–589.

[26] M. Clerc, J. Kennedy, *The Particle Swarm – Explosion, Stability, and Convergence In a Multidimensional Complex Space*, *IEEE Transactions on Evolutionary Computation,* 6, 2002, 58–73.

[27] J. Kennedy, *Small Worlds and Mega-Minds: Effects of Neighborhood Topology on Particle Swarm Performance*, *Proceedings of the IEEE Congress on Evolutionary Computation*, 1999, 1931–1938.

[28] S. Janson, M. Middendorf, *A Hierarchical Particle Swarm Optimizer and Its Adaptive Variant, IEEE Transactions on System Man and Cybernetics B.*, 35, 2005, 1272–1282.

[29] R. Salomon, *Reevaluating Genetic Algorithm Performance Under Coordinate Rotation of Benchmark Functions*, *BioSystems,* 39, 1996, 263–278.

[30] J. Kennedy, *Bare bones particle swarms*, *Proceedings of the IEEE Swarm Intelligence Symposium*, 2003, 80–87.

[31] H. Zhang, D. D. Kennedy, G. P. Rangaiah, A. Bonilla-Petriciolet, *Novel bare-bones particle swarm optimization and its performance for modeling vapor-liquid equilibrium data, Fluid Phase Equilibria*, 301, 2011, 33-45.

In: Global Optimization
Editor: Angelika Michalski
ISBN: 978-1-62417-796-5

Chapter 2

PARTICLE SWARM GLOBAL OPTIMIZATION OF ORBITAL MANEUVERS

*Mauro Pontani**
Scuola di Ingegneria Aerospaziale,
University "La Sapienza", Rome, Italy

ABSTRACT

The particle swarm optimization algorithm is a population-based stochastic method developed in recent years and successfully applied in several fields of research, ranging from chemical engineering to computer science. It represents a very intuitive methodology for global optimization, inspired by the behavior of bird flocks while searching for food. The particle swarm optimization technique takes advantage of the mechanism of information sharing that affects the overall behavior of a swarm, with the intent of determining the optimal values of the unknown parameters of the problem under consideration. This work describes the application of the method to a variety of space trajectory optimization problems that exhibit multiple local optimal solutions. Specifically, the swarming algorithm is applied to: (i) transfer trajectories between two coplanar circular orbits; (ii) two-impulse transfer between two non-coplanar elliptic orbits; (iii) four-impulse rendezvous trajectory between a pursuer spacecraft and a target space vehicle; (iv) two impulse transfer from a low Earth orbit to a low Moon orbit. Despite its simplicity and

* Email: mauro.pontani@uniroma1.it

intuitiveness, the particle swarm algorithm proves to be quite effective in finding the globally optimal solution to all of the applications considered in the chapter, with great numerical accuracy.

Keywords: Swarming theory, Heuristic optimization methods, Globally optimal orbital maneuvers, Impulsive orbital transfers and rendezvous, Earth-Moon missions, PSO Particle Swarm Optimization, GA Genetic Algorithms, SA Simulated Annealing, P Pursuer spacecraft, T Target spacecraft

1. Introduction

The determination of optimal (either minimum-time or minimum-propellant-consumption) space trajectories has been pursued with different numerical optimization methods, for decades. In general, numerical optimization methods can be classified as deterministic or heuristic methods. Deterministic gradient-based methods assume the continuity and differentiability of the objective function to be minimized. In addition, gradient-based methods are local in nature and require the identification of a suitable first attempt "solution" in the region of convergence, which is unknown *a priori* and strongly problem dependent. These circumstances have motivated the development of effective heuristic methods in the last decades. These algorithms are also referred to as evolutionary algorithms, and are inspired by natural phenomena. Evolutionary computation techniques exploit a population of individuals, representing possible solutions to the problem of interest.

The optimal solution is sought through cooperation and competition among individuals. The most popular class of these techniques is represented by the genetic algorithms (GA) [1], which model the evolution of a species, based on Darwin's principle of survival of the fittest. Differential evolution algorithms represent alternative stochastic approaches with some analogy with genetic algorithms, in the sense that new individuals are generated from old individuals and are eventually preserved after comparing them with their parents. Ant colony optimization [2] is another method, inspired by the behavior of ants, whereas the simulated annealing algorithm (SA) [2] mimics the equilibrium of large numbers of atoms during an annealing process.

The particle swarm optimization (PSO) technique, which is the methodology being addressed in this work, was first introduced by Eberhart and Kennedy [3,4] in 1995 and belongs to the category of swarm intelligence methods [2,5]. It mimics the unpredictable motion of bird flocks while searching for food, taking advantage of the mechanism of information sharing that affects the overall behavior of a swarm [2-16]. The initial population that composes the swarm is randomly generated at the first iteration of the process. Each particle is associated with a position vector and a velocity vector at a given iteration. More specifically, the position vector includes the values of the unknown parameters of the problem, whereas the velocity vector determines the position update. Each particle represents a possible solution to the problem, and corresponds to a specific value of the objective function. At the end of the process, the best particle (i.e. the best solution with reference to the objective function) is selected. Both the position and the velocity vector are updated in a single iteration. For each particle the formula for velocity update includes three terms with stochastic weights; one of these terms is the so-called social component, related to the collective best position ever visited by a portion of the particles that form the swarm.

A number of options are available for implementing the PSO technique. First of all, different values for the stochastic weights have been proposed [10-12], and they seem to affect both the algorithm convergence and the capability to detect the global optimum. In addition, two different versions of the particle swarm exist [2,5]: the *global* version, where the collective best position (associated with the social term in the velocity updating expression) is selected by considering the entire swarm, and the *local* version, where for each particle the collective best position is selected among the particles located in a proper neighborhood of the particle itself. Albeit less computationally efficient, in the scientific literature [2,5,7-11] the local version of the PSO algorithm, based on the definition of the neighbors of each particle, has been reported to be occasionally capable of avoiding local minima. Additional improvements based on the application of evolutionary operators to the particle swarm methodology are also reported by several researchers [17,18].

The basic version of the particle swarm algorithm appears as very intuitive and is extremely easy to program. In addition, this kind of method is well suited for finding the globally optimal solution to an optimization problem, and requires only the definition of the search space for the unknown parameters. Although computationally expensive with respect to gradient-based methods [13], in the scientific literature [19-21] the PSO technique is reported to be more efficient when compared to GA, due to a reduced number of function evaluations.

Despite its promising features and the vast number of papers devoted to this technique, most researchers concentrated on topological [22] and multimodal mathematical problems [15,23], and only a limited number of applications appear of practical interest. In the scientific literature several papers describe the use of the particle swarm methodology in the context of chemical processes [24,25]. Fourie and Groenwold [26] employed the PSO technique for shape and size optimization in structural engineering. Most recently Khurana et al. [27] applied the swarming theory to airfoil shape optimization. Bessette and Spencer [28,29] successfully employed the PSO technique for space trajectory optimization, focusing on the optimization of time-limited orbital transfers [28] and impulsive interplanetary trajectories [29].

In both papers they claim that the PSO method outperforms the DE algorithm, with regard to convergence speed as well as to reliability. Vasile et al. [30] investigated several algorithms for global optimization (including PSO and DE), employing different settings for each of them and using the success rate as the main index for the performance evaluation of each method. In their paper an improved algorithm based on differential evolution is proposed. Nevertheless in the conclusions the authors remark the importance of appropriately coupling the evolutionary technique with the problem structure, stating that the optimal choice of a method of solution is definitely problem dependent. Zhu et al. [31,32] combined DE and PSO for satellite constellation design [31] and for determining globally optimal low-thrust trajectories for asteroid exploration [32]. Rosa Sentinella and Casalino [33] proposed a hybrid optimization procedure that runs three different optimizers – based on GA, DE, and PSO – in parallel.

They applied this method to the optimization of multiple-impulse rendezvous trajectories and of Earth-to-Mars round trip missions. Most recently, Pontani and Conway [34-39] and Pontani et al. [40,41] employed the PSO for the determination of Lyapunov and Halo (periodic) orbits in the circular restricted three-body problem and for the optimization of impulsive and finite-thrust trajectories (using both the nonlinear dynamics equations and the linear Hill-Clohessy-Wiltshire equations). The work that follows adopts a basic, general purpose global version of the PSO method, according to the general settings suggested in Refs. 6-8. A simple MATLAB code has been implemented and applied to a variety of space trajectory optimization problems that exhibit multiple local minima.

In particular, the following applications have been considered:

(1) *Optimal two- and three-impulse orbital transfer between coplanar circular orbits.* The problem consists of determining the optimal locations, directions, magnitudes, and number of the impulsive changes of velocity that allow transferring a spacecraft from an initial circular orbit to a final circular orbit (while minimizing the characteristic velocity of the maneuver, i.e. propellant consumption). Two well known possible global optimal transfers exist: the (two-impulse) Hohmann transfer and the (three-impulse) bielliptic transfer. The final choice between the two options depends on the radii of the terminal orbits, as well as on the maximum apoapse radius of the intermediate arc of the trajectory.

(2) *Optimal two-impulse orbital transfer between two non-coplanar elliptic orbits.* The problem consists of determining the optimal locations, directions, and magnitudes of the velocity impulses that allow transferring a spacecraft from an initial elliptic orbit to a final (non-coplanar) elliptic orbit – with arbitrary mutual orientation – while minimizing the characteristic velocity of the maneuver. This kind of problem exhibits multiple minima, which are very challenging to discriminate with classical (deterministic) optimization methods.

(3) *Optimal four-impulse rendezvous.* The problem consists of determining the optimal locations, directions, and magnitudes of the four impulsive changes of velocity that allow performing a coplanar circle-to-circle rendezvous between a target vehicle and a pursuer spacecraft, while minimizing again the characteristic velocity of the overall orbital maneuver. For this case also multiple local minima exist and the determination of the globally optimal solution with classical (deterministic) methods is quite challenging.

(4) *Optimal two-impulse transfer from low Earth orbit to low Moon orbit.* The problem consists of determining the optimal locations, directions, and magnitudes of the two impulsive changes of velocity that allow transferring a spacecraft from an initial low Earth orbit to a final low Moon orbit – with specified perigee and periselenium radii – while minimizing the characteristic velocity of the maneuver, also referred to as delta-v budget.

The work that follows is intended to point out that the particle swarm algorithm, even in its simplest formulation and without interacting with other algorithms, is capable of effectively solving a variety of multimodal space trajectory optimization problems with great numerical accuracy.

2. Description of the Method

In the context of space trajectories the optimization problems of interest usually consist of minimizing a given objective function related with the time evolution of a dynamical system, which can be governed either by differential equations or by algebraic equations. The minimization is achieved by selecting the optimal values of the unknown parameters and time-varying variables. If the system dynamics are governed by a set of algebraic (nonlinear) equations the problem reduces to a nonlinear programming problem and the optimization process is aimed at finding the optimal values of a set of unknown parameters.

2.1. Unconstrained Optimization

Unconstrained parameter optimization problems can be stated as follows: determine the optimal values of the n unknown parameters $\{\chi_1,\ldots,\chi_n\}$ such that the objective function J is minimized. The time evolution of the dynamical system under consideration depends on $\{\chi_1,\ldots,\chi_n\}$, which are constrained to their respective ranges

$$a_k \le \chi_k \le b_k \qquad (k=1,\ldots,n) \tag{1}$$

As mentioned above, the PSO technique is a population-based method, where the population is represented by a swarm of N particles. Each particle i $(i=1,\ldots,N)$ is associated with a position vector $\boldsymbol{\chi}(i)$ and with a velocity vector $\boldsymbol{w}(i)$. The position vector includes the values of the n unknown parameters of the problem,

$$\boldsymbol{\chi}(i)=\begin{bmatrix}\chi_1(i) & \ldots & \chi_n(i)\end{bmatrix}^T \tag{2}$$

whereas the velocity vector, whose components are denoted with $w_k(i)$ $(k=1,\ldots,n)$, determines the position update (both $\boldsymbol{\chi}$ and $\boldsymbol{w}$ are defined as n-dimensional column vectors). As the position components are bounded, also the corresponding velocity components must be constrained to suitable ranges,

$$-(b_k - a_k) \le w_k \le (b_k - a_k) \quad \Rightarrow \quad -d_k \le w_k \le d_k \tag{3}$$

where $d_k = b_k - a_k \quad (k = 1, \ldots, n)$. The limitations (3) are due to the fact that if $w_k > b_k - a_k$ or $w_k < a_k - b_k$ then, starting from any coordinate $\chi_k^{(j)}$ (at iteration *j*), the updated coordinate $\chi_k^{(j+1)}$ would violate the condition (1). If $\boldsymbol{a} = [a_1 \quad \ldots \quad a_n]^T$, $\boldsymbol{b} = [b_1 \quad \ldots \quad b_n]^T$, and $\boldsymbol{d} = [d_1 \quad \ldots \quad d_n]^T$, the relationships (1) and (2) can be rewritten in compact form as

$$\boldsymbol{a} \le \boldsymbol{\chi} \le \boldsymbol{b} \quad \text{and} \quad -\boldsymbol{d} \le \boldsymbol{w} \le \boldsymbol{d} \tag{4}$$

The initial population for the PSO algorithm is randomly generated by introducing *N* particles, whose positions and velocities are (stochastically) uniformly distributed in the respective search spaces, defined by Eq. (4). Each particle represents a possible solution to the problem, and corresponds to a specific value of the objective function. The expressions for position and velocity update determine the swarm evolution toward the location of the globally optimal position, which corresponds to the globally optimal solution to the problem under consideration. The following steps compose the generic iteration *j*:

a) for $i = 1, \ldots, N$:
   (i) evaluate the objective function associated with particle *i*, $J^{(j)}(i)$
   (ii) determine the best position ever visited by particle *i* up to the current iteration *j*, $\boldsymbol{\psi}^{(j)}(i)$: $\boldsymbol{\psi}^{(j)}(i) = \boldsymbol{\chi}^{(l)}(i)$, where $l = \arg\min_{p=1,\ldots,j} J^{(p)}(i)$

b) determine the global best position ever visited by the entire swarm, $\boldsymbol{Y}^{(j)}$: $\boldsymbol{Y}^{(j)} = \boldsymbol{\psi}^{(j)}(q)$, where $q = \arg\min_{i=1,\ldots,N} \Im^{(j)}(i)$ and

$\Im^{(j)}(i)$ $(i=1,\ldots,N)$ represents the value of the objective function corresponding to the best position ever visited by particle *i* up to iteration *j*, i.e. $\Im^{(j)}(i)=\min_{p=1,\ldots,j} J^{(p)}(i)$

c) update the velocity vector. For each particle *i* and for each component $w_k(i)$ $(k=1,\ldots,n;\ i=1,\ldots,N)$,

$$w_k^{(j+1)}(i)=c_I w_k^{(j)}(i)+c_C\left[\psi_k^{(j)}(i)-x_k^{(j)}(i)\right]+c_S\left[Y_k^{(j)}-x_k^{(j)}(i)\right] \tag{5}$$

The *inertial*, *cognitive*, and *social* (stochastic) weights have the following expressions [6-8]:

$$c_I=\frac{1+r_1(0,1)}{2} \qquad c_C=1.49445\,r_2(0,1) \qquad c_S=1.49445\,r_3(0,1) \tag{6}$$

where $r_1(0,1)$, $r_2(0,1)$, and $r_3(0,1)$ represent three independent random numbers with uniform distribution between 0 and 1. Then,

(i) if $w_k^{(j+1)}(i)<-d_k \;\Rightarrow\; w_k^{(j+1)}(i)=-d_k$

(ii) if $w_k^{(j+1)}(i)>d_k \;\Rightarrow\; w_k^{(j+1)}(i)=d_k$

d) update the position vector. For each particle *i* and for each component $\chi_k(i)$ $(k=1,\ldots,n;\ i=1,\ldots,N)$,

$$\chi_k^{(j+1)}(i)=\chi_k^{(j)}(i)+w_k^{(j)}(i) \tag{7}$$

(i) if $\chi_k^{(j+1)}(i)<a_k \;\Rightarrow\; \chi_k^{(j+1)}(i)=a_k$ and $w_k^{(j+1)}(i)=0$

(ii) if $\chi_k^{(j+1)}(i)>b_k \;\Rightarrow\; \chi_k^{(j+1)}(i)=b_k$ and $w_k^{(j+1)}(i)=0$

The algorithm terminates when the maximum number of iterations N_{IT} is reached. The position vector of the global best particle, $\boldsymbol{Y}^{(N_{IT})}$, is expected to contain the optimal values of the unknown parameters, which correspond to the global minimum of *J*, denoted with $J_{opt}^{(N_{IT})}$. A fairly large number of iterations are set to ensure that the final solution is stable enough to be considered optimal. The central idea underlying the method is contained in the formula (5) for velocity updating. This formula includes three terms with stochastic weights: the first term is the so called *inertial* component and for each particle is proportional to its velocity in the preceding iteration; the second component is termed the *cognitive* component, directed toward the personal best position, i.e. the best position experienced by the particle; and finally the third term is the *social* component, directed toward the global best position, i.e. the best position yet located by any particle in the swarm. According to point (c), if a component w_k of the velocity vector violates Eq. (3), then w_k is set to the minimum (maximum) value $-d_k$ (d_k). If a component χ_k of the position vector violates Eq. (1), then χ_k is set to the minimum (maximum) value a_k (b_k), and the corresponding velocity component is set to 0. This ensures that in the successive iteration the update of the velocity component is not affected by the first term of Eq. (5), which could lead the particle to again violating the constraint (1).

2.2. Constrained Optimization

Space trajectory optimization problems must be frequently modeled as constrained optimization problems, i.e. they involve equalities and/or inequalities, regarding (directly or indirectly) the unknown parameters. The PSO algorithm described in the preceding subsection must be suitably adjusted to deal with constrained problems.

In general, evolutionary computation methods encounter difficulties in treating equality constraints [6-9,14,42], because they narrow considerably the search space where feasible solutions can be located. This is due to the fact that (non-redundant) equality constraints actually reduce the degree of freedom of the problem according to their number. However, in the problems considered in this chapter only inequality constraints are involved. They are less problematic because they reduce the search space of feasible solutions without decreasing the degree of freedom of the problem. For each particle the simplest way of treating inequality constraints consists of assigning a fictitious infinite value to the objective function if the particle violates at least one of the constraints. In addition, the corresponding velocity is set to zero, so that the successive velocity update – according to Eq. (5) – is affected only by the social term and by the cognitive term. This circumstance statistically leads the particle to a feasible region of the search space. Point (a)-(i) of the algorithm (in the preceding subsection) must be modified as follows:

a) for $i = 1, \ldots, N$
   (i) evaluate the inequality constraints. If (at least) one of them is violated *then* set $J^{(j)}(i) = \infty$ and $\boldsymbol{w}^{(j+1)}(i) = \boldsymbol{0}$ and skip point (c), *else* evaluate the objective function associated with particle *i*, $J^{(j)}(i)$

Further options exist for dealing with constraints (e.g., methods based on preserving feasibility of solutions, methods that distinguish between feasible and infeasible solutions, cf. Refs. [14, 42-45]). However, the simple strategy described above for inequality constraints is quite effective for the problems considered in this chapter.

3. Optimal Impulsive Transfers between Coplanar Circular Orbits

The impulsive thrust approximation has application in the case of high thrust rockets and spacecraft and represents a very good approximation [46,47] of the finite-thrust maneuvers that a vehicle performs once it is in orbit, under the assumption that gravitational losses during propulsion can be neglected. The globally optimal impulsive transfer between two circular orbits is composed of either two or three impulses (Hohmann or bielliptic transfer), which must occur at the apse points. Several authors [48-53] demonstrated analytically this result in the last decades. If the terminal orbits are circular (with R_1 denoting the radius of the initial orbit and R_2 the radius of the final orbit), the condition $0.08376 < R_2/R_1 < 11.939$ ensures unequivocally that the globally optimal transfer is the bitangent Hohmann transfer, whereas the conditions $R_2/R_1 < 0.06418$ or $R_2/R_1 > 15.582$ imply the global optimality of the bielliptic transfer, regardless of the maximum allowed apoapse radius, at which the midcourse impulse occurs. What will be demonstrated is that the PSO method is capable of numerically determining the globally optimal (two- or three-) impulse transfer in each case.

3.1. Problem Definition

The problem consists of determining the optimal locations, directions, magnitudes, and number of the impulses that allow transferring a spacecraft from a circular orbit of radius R_1 to a circular orbit of radius R_2. Only external transfers are considered, and therefore $R_2 > R_1$. Theoretical results [49,53] state that as many as three impulses are needed for a globally optimal transfer between two circular orbits (in the absence of any time limitation). Hence, for the problem at hand the objective function, i.e. the characteristic velocity of the transfer, is represented by the sum of three terms,

$$J = \Delta v_1 + \Delta v_2 + \Delta v_3 \tag{8}$$

where Δv_1, Δv_2, and Δv_3 denote the magnitudes of the three impulsive changes of velocity.

The transfer trajectory is composed of two Keplerian arcs, which are assumed to be of elliptic type. The trajectory can be propagated in the coast arcs thorugh the relations that hold for Keplerian motion, as shown in Ref. 39. It is relatively straightforward [39] to prove that the following five parameters suffice to determine unequivocally a three-impulse orbital transfer between two coplanar circular orbits:

$$\{\Delta v_1, \delta_1, \Delta v_2, \delta_2, \Delta f_1\} \tag{9}$$

where Δv_i and δ_i $(i = 1, 2)$ are the magnitude and direction (with respect to the local horizontal) of the velocity impulse *i*, whereas Δf_1 is the true anomaly variation during the first coast arc. In addition, the transfer trajectory is assumed to be subject to a constraint on the radius of greatest recession from the attracting body. This condition implies that the radius must not exceed a limiting value, denoted with R_M. It is worth remarking that with the present formulation a two-impulse transfer represents a special solution that corresponds to the vanishing of one of the three impulses.

3.2. Numerical Results

For the problem at hand the PSO algorithm employs 100 particles $(N=100)$ and is run for 500 iterations $(N_{IT}=500)$. Each particle includes the values of the five unknown parameters (9).

The problem is solved by employing a normalized set of units: the radius of the initial orbit represents the distance unit (DU), whereas the time unit (TU) is such that $\mu=1\ \mathrm{DU}^3/\mathrm{TU}^2$. The search space is defined by the following inequalities:

$$0\ \frac{\mathrm{DU}}{\mathrm{TU}}\le \Delta v_i \le 1\ \frac{\mathrm{DU}}{\mathrm{TU}} \qquad 0\le \delta_i \le 2\pi \qquad 0\le \Delta f_1 \le 2\pi \qquad (i=1,2) \tag{10}$$

Two orbital transfers are considered as illustrative examples, corresponding to two different radii of the terminal orbit: (a) $R_2=10$ DU and (b) $R_2=20$ DU. The maximum allowed distance from the attractive body, R_M, is set to 100 DU. For case (a) the globally optimal transfer is represented by the Hohmann transfer, whereas for case (b) the bielliptic transfer is globally optimal.

With regard to case (a) the algorithm is able to detect with great accuracy the globally optimal solution, which corresponds to applying two impulsive changes of velocity at the apse points of the intermediate arc of the trajectory. Ten runs of the algorithm were performed and the results yielded by the PSO algorithm are virtually indistinguishable from each other. If J_H^* denotes the minimum characteristic velocity calculated with the analytical formulas of the Hohmann transfer and $J_{opt}^{(N_{IT})}$ is the optimal value of J obtained by the PSO method, $\left(J_{opt}^{(N_{IT})}-J_H^*\right)$ represents the displacement of the objective function found by the PSO method from the respective theoretical optimal value. For all of the runs the displacements $\left(J_{opt}^{(j)}-J_H^*\right)$ do not exceed 10^{-16}, and this points out that the final numerical results are virtually indistinguishable from those obtained through analysis. Figure 1(a) illustrates the globally optimal Hohmann transfer for the case at hand.

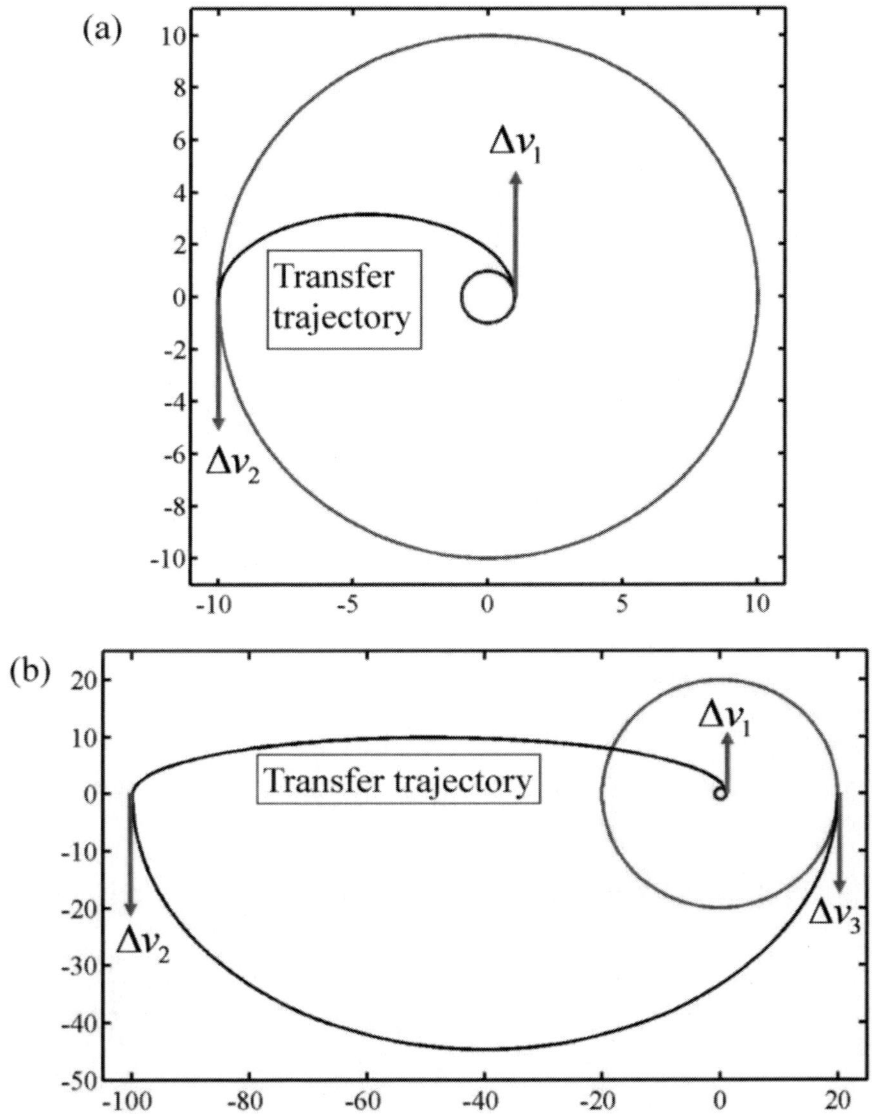

Figure 1. Hohmann (a) and bielliptic (b) globally optimal transfers.

For case (b) the number of iterations was increased to 1000 in order to obtain more refined results. Ten runs of the algorithm were performed with

100 particles and in 9 out of 10 cases the algorithm prematurely converged to the locally optimal solution corresponding to the Hohmann transfer (with $J_H^* = 0.534731$ DU/TU). This is mainly due to the inequality constraints (related to the radius of greatest recession from the attracting body), which drastically reduce the number of particles associated with feasible trajectories, thus diminishing the search space exploration capability of the algorithm. Improved results can be obtained by employing a greater number of particles. If J_B^* $(= 0.515927 \text{ DU/TU})$ denotes the minimum characteristic velocity calculated with the analytical formulas of the bielliptic transfer, $\left(J_{opt}^{(N_{IT})} - J_B^*\right)$ represents the displacement of the objective function found by the PSO method from the respective theoretical optimal value. By definition, the algorithm converges to the globally optimal solution if $\left(J_{opt}^{(N_{IT})} - J_B^*\right) < 10^{-3}$ (in DU/TU). Table 1 reports the success rate (over 10 runs) of the PSO algorithm when 100, 300, 500, and 1000 particles are employed, as well as the corresponding best values of $\left(J_{opt}^{(N_{IT})} - J_B^*\right)$.

As expected, the success rate increases as the population size increases. Even employing 100 particles the best result attained with the PSO method is extremely close to the theoretical global optimal solution. However, since in this case this good result is only found for 1 run out of 10 (but in every run when 1000 particles are used) this simple example illustrates that with no *a priori* information, the largest practical population should be used on a new problem. Figure 1(b) illustrates the globally optimal bielliptic transfer for the case at hand.

Table 1. Optimal three-impulse transfer: success rate and best results for different numbers of particles

N	100	300	500	1000
Success rate	10 %	40 %	50 %	100 %
$\min\left(J_{opt}^{(N_{IT})} - J_B^*\right)$	$4.647 \cdot 10^{-5}$	$6.817 \cdot 10^{-6}$	$1.201 \cdot 10^{-7}$	$2.119 \cdot 10^{-9}$

4. Optimal Two-Impulse Transfers between Non-Coplanar Orbits

Unlike the case of coplanar orbital transfers between two circular orbits, the problem of transferring a spacecraft from an initial elliptic orbit to a final, non-coplanar elliptic orbit (employing velocity impulses) is not amenable to an analytical solution. Yet, recently Stratemeier [54] has shown that multiple minima exist for the problem of interest, which consists of determining the optimal locations, magnitudes, and directions of the two impulsive changes of velocity that allow performing the transfer.

4.1. Problem Definition

The two-impulse globally optimal transfer between two elliptic non-coplanar orbits minimizes the characteristic velocity of the orbital maneuver, and therefore the objective function for the problem of interest is

$$J = \Delta v_1 + \Delta v_2 \tag{11}$$

where Δv_1 and Δv_2 represent the magnitudes of the two velocity impulses at the departure point and at the arrival point, respectively. The initial and the final orbit (associated with the subscripts "0" and "f") are defined through their respective orbital elements [52], i.e. semi-major axis a_j, eccentricity e_j, inclination i_j, right ascension of the ascending node (RAAN) Ω_j, and argument of periapse ω_j $\left(j = 0 \text{ or } f\right)$. Let f_0 and f_f denote respectively the true anomaly on the initial orbit at the departure point (where the first impulse is applied) and the true anomaly on the final orbit at the arrival point (where the second impulse occurs). Immediately before the first impulse and after the second impulse both the position vector $\boldsymbol{r}$ and the velocity vector $\boldsymbol{v}$ are defined if f_0 and f_f are specified.

Due to the Lambert theorem [52] it can be proven [39] that the following three parameters suffice to determine unequivocally an orbital transfer between two elliptic non-coplanar orbits:

$$\left\{f_0, f_f, \Delta t\right\} \tag{12}$$

where $\Delta t \left(= t_f - t_0\right)$ is the transfer time.

4.2. Numerical Results

For the problem at hand the PSO algorithm employs 50 particles $(N = 50)$ and is run for 3000 iterations $(N_{IT} = 3000)$. Each particle includes the values of the three unknown parameters (12).

The problem is solved by employing a normalized set of units: the Earth radius represents the distance unit (DU), whereas the time unit (TU) is such that $\mu_E = 1\ \mathrm{DU}^3/\mathrm{TU}^2$ (and therefore $1\ \mathrm{DU} = 6378.165\ \mathrm{km}$ and $1\ \mathrm{TU} = 806.8\ \mathrm{sec}$). The unknown parameters are constrained to the following ranges, which define the search space:

$$0 \le f_0 \le 2\pi \qquad 0 \le f_f \le 2\pi \qquad 0\ \mathrm{TU} \le \Delta t \le 20\ \mathrm{TU} \tag{13}$$

The initial and the final orbit are defined through their respective orbital elements, whose values are the same as those employed by Stratemeier [54] and are reported in Table 2.

Tables 2-4 collect the results of the optimization process, i.e. the optimal locations, components, and magnitudes of the two impulses associated with the globally optimal transfer trajectory, as well as the values of its orbital elements. The components are referred to the Earth-centered inertial (ECI) frame [52].

Figure 2 portrays the globally optimal transfer trajectory, corresponding to a minimum value of J equal to $1.392969\ \mathrm{km/sec}$. Stratemeier [54] identified four minima for the problem at hand. The global optimal transfer obtained with the PSO method is virtually indistinguishable from the global optimal solution found by Stratemeier, with regard both to the objective function $\left(J = 1.39296\ \mathrm{km/sec}\right)$ and to the optimal values of the unknown parameters $\left\{f_0, f_f, \Delta t\right\}$.

Table 2. Two-impulse transfer: orbital elements of the initial and final orbits, and of the transfer arc $(j=0 \text{ or } f \text{ or } T)$

	a_j (km)	e_j	i_j (deg)	Ω_j (deg)	ω_j (deg)
Initial orbit	9645.833	0.2	5.0	0.0	270.0
Final orbit	11575.0	0.2	0.0	0.0	30.0
Transfer arc	12439.447	0.104	5.25	7.54	21.07

Table 3. Two-impulse transfer: true anomalies and velocity impulses associated with the globally optimal solution

	f_0 (deg)	f_f (deg)	Δv_j (km/sec)
1st impulse	163.71	/	0.814379
2nd impulse	/	157.54	0.578590

Table 4. Two-impulse transfer: components of the globally optimal velocity impulses $(j=1,2)$

	Δv_{jX} (km/sec)	Δv_{jY} (km/sec)	Δv_{jZ} (km/sec)
1st impulse	–0.761707	0.270087	0.100340
2nd impulse	–0.272529	0.199610	0.469735

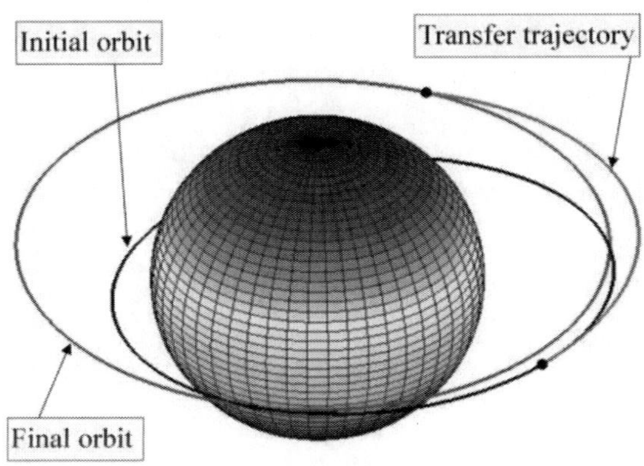

Figure 2. Globally optimal transfer trajectory between two noncoplanar elliptic orbits.

5. Optimal Four-Impulse Orbital Rendezvous

The orbital rendezvous of a pursuer spacecraft with a target vehicle (respectively denoted with "P" and "T" henceforward) in a specified time can be achieved by two (or more) impulses. If the orbits of the two spacecraft are coplanar and the time is sufficiently large, the globally optimal (two-impulse) solution is the Hohmann transfer. More precisely, the Hohmann transfer represents the optimal rendezvous if the specified transfer time, denoted with Δt_T, is not less than the sum of two terms: (a) the time needed for a correct phasing between P and T, Δt_P; (b) the Hohmann transfer time between the specified terminal orbits, Δt_H. This means that the Hohmann transfer is the (globally) optimal two-impulse rendezvous if

$$\Delta t_T \geq \Delta t_P + \Delta t_H \tag{14}$$

If the terminal orbits are circular with radii R_P and R_T, if $0.08376 \leq R_T / R_P \leq 11.939$, and the condition (14) holds, then the Hohmann transfer is the globally optimal rendezvous without any limitation on the number of impulses. For those cases for which the condition (14) is violated additional impulses can allow performing the rendezvous with substantial propellant savings with respect to the use of two impulses. Analytical results exist for linearized coplanar circle-to-circle rendezvous. In fact, Prussing [55,56] proved that as many as four impulses are required for an optimal time-fixed rendezvous. Analytical solutions arising from the linear theory can be employed for locating the optimal rendezvous also when the nonlinear equations of motion are employed [57]. Yet, for the four-impulse rendezvous that is being considered in this section the extension of the linear theory to the nonlinear case only yielded a locally optimal solution [57], which can be outperformed by a completely different four-impulse rendezvous trajectory, as shown by Colasurdo and Pastrone [58].

5.1. Problem Definition

The problem consists of determining the optimal locations, directions, and magnitudes $\left\{\Delta v_i\right\}_{i=1,2,3,4}$ of the four velocity impulses that allow performing the orbital rendezvous between the pursuer spacecraft P and the target vehicle T in a specified time. The spacecraft P and T are placed in two circular orbits of radii R_P and R_T, respectively. The initial angular displacement between T and P is denoted with $\Delta\xi$ ($\Delta\xi > 0$ if T is located ahead with respect to P).

The globally optimal rendezvous, which is sought by the PSO algorithm, minimizes the delta-v budget of the overall orbital maneuver. This means that the objective function for the problem at hand is

$$J = \Delta v_1 + \Delta v_2 + \Delta v_3 + \Delta v_4 \tag{15}$$

It is worth noticing that two- and three-impulse trajectories correspond to the vanishing of one or two impulses and can be detected by the algorithm, provided that they represent convenient solutions (in terms of characteristic velocity).

It can be demonstrated [40] that the following six parameters define the rendezvous trajectory:

$$\{\Delta v_1, \delta_1, \Delta v_2, \delta_2, \Delta E_1, \Delta E_2\} \tag{16}$$

where Δv_1 and Δv_2 represent the magnitudes of the first two impulsive changes of velocity, δ_1 and δ_2 their respective directions (relative to the local horizontal), and finally ΔE_1 and ΔE_2 denote the eccentric anomaly variations during the first two Keplerian arcs of the trajectory.

5.2. Numerical Results

For the problem at hand the PSO algorithm employs 100 particles $(N = 100)$ and is run for 2000 iterations $(N_{IT} = 2000)$. Each particle includes the values of the six unknown parameters (16).

The problem is solved by employing a normalized set of units: the radius of the initial orbit represents the distance unit (DU), whereas the time unit (TU) is such that $\mu = 1\ \mathrm{DU}^3/\mathrm{TU}^2$. The search space is defined by the following inequalities:

$$0\ \frac{\text{DU}}{\text{TU}} \le \Delta v_i \le 0.2\ \frac{\text{DU}}{\text{TU}} \qquad 0 \le \delta_i \le 2\pi \qquad 0 \le \Delta E_i \le 4\pi \qquad (i = 1,2) \tag{17}$$

The following illustrative example coming from the scientific literature [57] is considered:

$$R_P = 1\ \text{DU} \qquad R_T = 1\ \text{DU} \qquad \Delta\xi = \pi \qquad t_f = 4.6\pi\ \text{TU}$$

Both spacecraft are placed in the same circular orbit of radius $R_P\ (= R_T)$, and their angular displacement at the initial time is $\Delta\xi = \pi$. In this case the time needed for a correct phasing for a Hohmann transfer tends to infinity $(\Delta t_P \to \infty)$, and therefore the Hohmann transfer cannot represent the optimal rendezvous trajectory, regardless of the transfer time.

Tables 5-6 collect the results of the optimization process, i.e. the optimal locations, directions, and magnitudes of the four impulses associated with the (globally) optimal rendezvous trajectory for each case, the radii *r* (DU) at which the impulses are applied, and the values of the semi-major axes and eccentricities of the three Keplerian arcs that compose the optimal trajectory. Figure 3 portrays the optimal rendezvous trajectory, corresponding to the following minimum value of *J*:

$$J_{opt}^{(N_{IT})} = 0.163828\ \text{DU/TU}$$

The solution obtained with the PSO method is virtually indistinguishable from that found by Colasurdo and Pastrone [58]. The numerical results exhibit a high degree of symmetry. In fact, from Tables 5 and 6 it is apparent that

$$\begin{gathered}
\Delta v_1 \approx \Delta v_4 \qquad \Delta v_2 \approx \Delta v_3 \\
\text{mod}(\delta_1 - \delta_4, 2\pi) \approx \pi \qquad \text{mod}(\delta_2 + \delta_3, 2\pi) \approx \pi \\
r(t_1) \approx r(t_2) \qquad t_1 - t_0 \approx t_f - t_2 \\
a_1 \approx a_3 \qquad e_1 \approx e_3
\end{gathered} \tag{18}$$

Reference 40 contains the proof that for the problem at hand an optimal solution must be symmetric (i.e. that all of the relations (18) must hold precisely).

Hence, the fact that the relationships (18) are satisfied to a great accuracy (up to six digits) actually represents an additional proof of optimality of the solution found with the PSO methodology.

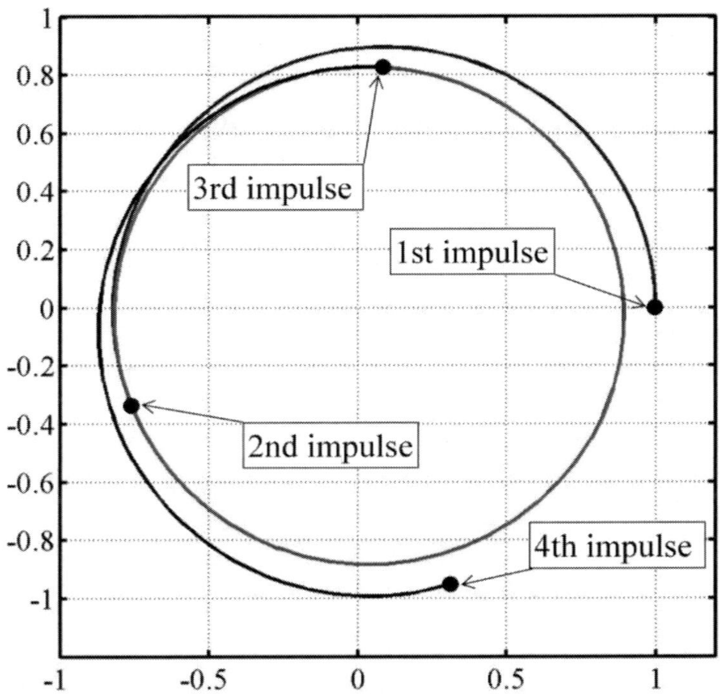

Figure 3. Optimal four-impulse rendezvous trajectory.

Table 5. Results for the optimal four-impulse rendezvous

Impulse	Δv (DU/TU)	δ (deg)	r (DU)	t (TU)
1	0.0511846	192.776	1	0
2	0.0307293	182.507	0.832237	2.996029
3	0.0307293	357.493	0.832237	11.455296
4	0.0511846	12.776	1	14.451326

Table 6. Four-impulse rendezvous: results for the Keplerian arcs of the trajectory

Keplerian arc	1	2	3
Semi-major axis a (DU)	0.911399	0.857262	0.911399
Eccentricity e	0.0979352	0.0540120	0.0979351

6. Optimal Low Earth Orbit – Low Moon Orbit Transfer

Earth-Moon orbit transfers have been investigated by many researchers in the last decades. Pioneering contributions are contained in Refs. 59-62. Most studies employ the planar circular restricted three-body problem [63] as a reasonably accurate model to investigate the orbital motion of a spacecraft subject to the simultaneous attraction of the Earth and the Moon. A variety of transfer options have been proposed, sometimes using space manifold dynamics [64-66] or low-thrust propulsion [67,68].

This section considers high-thrust transfers from a low Earth orbit (LEO) to a low Moon orbit (LMO), using the same dynamical framework and data reported in Ref. 69. Both LEO and LMO are assumed circular.

6.1. Problem Definition

In this study a simplified version of the circular restricted three-body problem is considered, in which the spacecraft, whose mass is negligible with respect to that of the Earth and the Moon, is subject to the gravitational attraction of the two celestial bodies (termed the primaries henceforward) at all times during the transfer. For the sake of simplicity, the following additional assumptions are introduced:

(i) Earth fixed in space;
(ii) Moon in circular orbit around the Earth ;
(iii) spacecraft trajectory entirely lying on the Moon orbital plane (2-d problem);

(iv) transfer trajectory including two high-thrust phases, modeled as impulsive changes of velocity, of magnitude Δv_{LEO} at the initial time t_0 (departure from LEO) and Δv_{LMO} at the final time t_f (injection in LMO);
(v) first impulse tangential to the initial circular orbit;
(vi) initial orbit of specified radius R_{LEO}, final orbit of specified radius;
(vii) R_{LMO} spacecraft subject to the only gravitational attraction of the Earth prior to departure from LEO and to the only gravitational attraction of the Moon after LMO injection

For the problem at hand, the equations of motion can be written in an inertial frame $(\hat{c}_1, \hat{c}_2, \hat{c}_3)$, with $\hat{c}_1$ aligned with the Moon position at t_0 and $\hat{c}_3$ directed along the Moon angular momentum (cf. Figure 4). As a result, the Moon orbit and the spacecraft trajectory (due to assumption (iii)) are both contained in the $(\hat{c}_1, \hat{c}_2)$-plane.

Let $\boldsymbol{r} = [X \quad Y][\hat{c}_1 \quad \hat{c}_2]^T$ and $\boldsymbol{v} = [V_X \quad V_Y][\hat{c}_1 \quad \hat{c}_2]^T$ denote the time-varying spacecraft position and velocity vectors. The equations of motion are

$$\dot{X} = V_X \tag{19}$$

$$\dot{Y} = V_Y \tag{20}$$

$$\dot{V}_X = -\frac{\mu_E X}{\left[X^2 + Y^2\right]^{3/2}} - \frac{\mu_M (X - X_M)}{\left[(X - X_M)^2 + (Y - Y_M)^2\right]^{3/2}} \tag{21}$$

$$\dot{V}_Y = -\frac{\mu_E Y}{\left[X^2 + Y^2\right]^{3/2}} - \frac{\mu_M (Y - Y_M)}{\left[(X - X_M)^2 + (Y - Y_M)^2\right]^{3/2}} \tag{22}$$

where μ_E $(=398600\ \text{km}^3/\text{sec}^2)$ and μ_M $(=4903\ \text{km}^3/\text{sec}^2)$ are respectively the Earth and Moon gravitational parameters, whereas X_M and Y_M represent the Moon position components in $(\hat{c}_1, \hat{c}_2)$. Due to assumption (ii), letting $t_0 = 0$,

$$X_M(t) = R_M \cos(\omega_M t) \quad \text{and} \quad Y_M(t) = R_M \sin(\omega_M t) \tag{23}$$

R_M $(=384400\ \text{km})$ and ω_M $\left(=\sqrt{\mu_E/R_M^3}\right)$ are the Moon orbit radius and rotation rate, respectively. Due to (23), the Moon inertial velocity components are given by

$$V_{XM}(t) = -\omega_M R_M \sin(\omega_M t) \quad \text{and} \quad V_{YM}(t) = \omega_M R_M \cos(\omega_M t) \tag{24}$$

The optimization problem consists of minimizing the characteristic velocity of the transfer, i.e.,

$$J = \Delta v_{LEO} + \Delta v_{LMO} \tag{25}$$

by choosing the optimal locations and magnitudes of both impulses and the optimal direction of the second impulse. However, what will be demonstrated is that the problem only depends on two parameters, after formulating it in a proper fashion.

To do this, the starting point is in establishing that the transfer trajectory is entirely determined if the initial (tangential) impulse location and magnitude are specified. In fact, the initial conditions for position and velocity at t_0^+ (immediately after the first impulse) determine the trajectory time evolution up to the possible encounter with the final desired lunar orbit. Under the assumption that this encounter occurs at point A (cf. Figure 4) at the time t_f, let $\boldsymbol{v}(t_f^-)$ denote the corresponding spacecraft velocity before applying the second impulse.

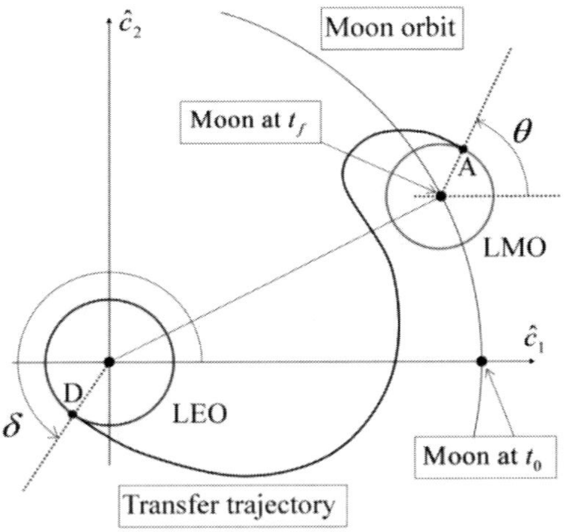

Figure 4. Geometry of the LEO-to-LMO orbital transfer.

The desired inertial velocity $\boldsymbol{v}\left(t_f^+\right)$ at A is expressed as the sum of the velocity in LMO (relative to the Moon) and the lunar orbital velocity (24) at t_f,

$$\boldsymbol{v}\left(t_f^+\right) = \sqrt{\frac{\mu_M}{R_{LMO}}}\left[\mp\sin\theta \quad \pm\cos\theta\right]\left[\hat{c}_1 \quad \hat{c}_2\right]^T + \omega_M R_M\left[-\sin\left(\omega_M t_f\right) \quad \cos\left(\omega_M t_f\right)\right]\left[\hat{c}_1 \quad \hat{c}_2\right]^T \tag{26}$$

where the two options ($\mp$ and $\pm$) correspond respectively to counterclockwise and clockwise orbits around the Moon, whereas the angle θ is illustrated in Figure 4 and is given by

$$\sin\theta = \frac{Y\left(t_f\right) - Y_M\left(t_f\right)}{R_{LMO}} \quad \text{and} \quad \cos\theta = \frac{X\left(t_f\right) - X_M\left(t_f\right)}{R_{LMO}} \tag{27}$$

Hence, the magnitude of the second velocity impulse is

$$\Delta v_{LMO} = \left|\boldsymbol{v}\left(t_f^+\right) - \boldsymbol{v}\left(t_f^-\right)\right| \tag{28}$$

As a result, the first velocity impulse Δv_{LEO} and the angle δ (portrayed in Figure 4) can be assumed as the only unknown parameters for the problem at hand. Of course, if the transfer trajectory does not intersect the final orbit, injection in LMO is infeasible and the second velocity impulse is not defined. This means that also the objective function (25) is not defined and this circumstance prevents any algorithm from converging (or even working). Hence, with the intent of circumventing this difficulty, the following alternative non-smooth objective function is defined:

$$J = \begin{cases} \Delta v_{LEO} + \Delta v_{LMO} & \text{if an intersection with LMO occurs at } t_f \in [t_0, t_{max}] \\ d_{min} & \text{if no intersection with LMO occurs in } [t_0, t_{max}] \end{cases} \quad (29)$$

d_{min} is the minimum distance of the spacecraft from the center of the Moon in the time interval $[t_0, t_{max}]$, and t_{max} represents a reasonable limiting value for trajectory integration. In this work t_{max} is set to 5 Earth sidereal days. It is worth remarking that in actual units the numerical value associated with d_{min} always exceeds the value of $(\Delta v_{LEO} + \Delta v_{LMO})$.

In summary, for each choice of the unknown parameters,

$$\{\Delta v_{LEO}, \delta\} \quad (30)$$

the spacecraft trajectory is integrated until either case (a) or (b) occurs: (a) interception with the final LMO occurs, (b) the time t_{max} has elapsed. Then, the objective is evaluated by means of (29).

6.2. Numerical Results

For the problem at hand the PSO algorithm employs 30 particles $(N = 30)$ and is run for 300 iterations $(N_{IT} = 300)$. Each particle includes the values of the two unknown parameters (30).

The problem is solved by employing actual units for the unknown parameters and for integrating the equations of motion. The search space is defined by the following inequalities:

$$3.025 \frac{\text{km}}{\text{sec}} \leq \Delta v_{LEO} \leq 3.162 \frac{\text{km}}{\text{sec}} \qquad 0 \leq \delta \leq 2\pi \tag{31}$$

The lower bound for Δv_{LEO} is the velocity value that corresponds to C_1, a special value of the Jacobi integral in the restricted three-body problem [63]. This value is associated with zero velocity curves containing the Lagrange point L_1 and is the inferior limiting value for an Earth-to-Moon transfer to be feasible. The upper bound for Δv_{LEO} is given by the escape velocity from the Earth gravitational field.

The initial LEO orbit has an altitude of 463 km, whereas three cases are considered for the final LMO, whose altitude is denoted with h_{LMO},

$$(1)\ h_{LMO} = 100 \text{ km} \qquad (2)\ h_{LMO} = 200 \text{ km} \qquad (3)\ h_{LMO} = 300 \text{ km}$$

Tables 7-8 collect the results of the optimization process, i.e. the magnitudes of the two impulses associated with the (globally) optimal LEO-to-LMO transfer for each case, as well as the optimal location of the first impulse (identified by δ) and the related transfer time t_f, for both counterclockwise and clockwise arrival at the Moon. From inspecting the results, it is apparent that the first velocity impulse is nearly independent of the final LMO, whereas the second impulse magnitude decreases as the LMO altitude increases. Moreover, the group of trajectories with counterclockwise arrival is slightly more economical (in terms of delta-v budget) with respect to the group of trajectories with clockwise arrival. The results obtained in this work are extremely close to those found by Miele and Mancuso [69] in terms of characteristic velocity (with displacements less than 1 m/sec for all of the cases), but differ with regard to the transfer times and the departure angle δ. This circumstance suggests that the objective functional (29) as a function of the unknown parameters is locally flat, i.e. extremely close values of the objective function correspond to relatively different values of the unknown parameters.

Table 7. LEO-to-LMO transfer with counterclockwise arrival at Moon: major parameters

h_{LMO} (km)	J (km/sec)	Δv_1 (km/sec)	Δv_2 (km/sec)	δ (deg)	t_f (days)
100	3.876	3.065	0.811	–118.68	4.41
200	3.861	3.065	0.796	–117.52	4.51
300	3.850	3.065	0.785	–113.58	4.87

(a)

(b)

(c)

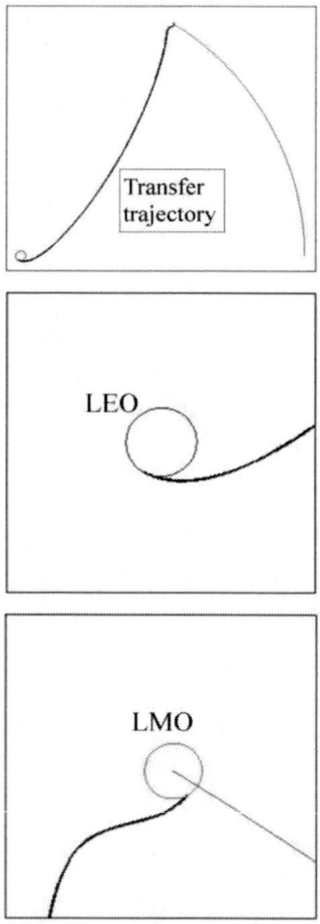

Figure 5. LEO-to-LMO transfer (counterclockwise arrival, $h_{LMO} = 100$ km): (a) overall trajectory, (b) trajectory arc in the proximity of the Earth, (c) trajectory arc in the proximity of the Moon.

Table 8. LEO-to-LMO transfer with clockwise arrival at Moon: major parameters

h_{LMO} (km)	J (km/sec)	Δv_1 (km/sec)	Δv_2 (km/sec)	δ (deg)	t_f (days)
100	3.881	3.068	0.813	−115.08	4.68
200	3.867	3.068	0.799	−114.10	4.77
300	3.854	3.068	0.786	−113.16	4.86

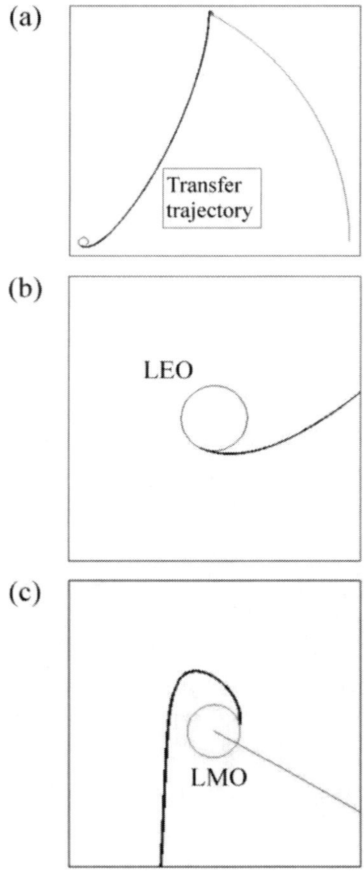

Figure 6. LEO-to-LMO transfer (clockwise arrival, $h_{LMO} = 100$ km): (a) overall trajectory, (b) trajectory arc in the proximity of the Earth, (c) trajectory arc in the proximity of the Moon.

With reference to case (1), Figures 5 and 6 depict the LEO-to-LMO transfer and the related trajectory arcs in the proximity of the Earth and the Moon, for both counterclockwise and clockwise arrival at Moon.

CONCLUSION

This chapter describes the use of an effective heuristic methodology, i.e. the particle swarm optimization technique, for optimizing space trajectories that exhibit multiple local optimal solutions. This method does not require continuity and differentiability of the objective function, does not need any guess to generate a solution, and is well suited for finding global optima. It is also based on very intuitive concepts, which make it very easy to program.

Four space trajectory optimization problems have been considered and a simple implementation of the PSO methodology has been successfully applied to their solution. Different numbers of particles and iterations have been employed, because the number of particles to use in each application depends on the problem complexity, in particular on the number of constraints and unknown parameters. A greater number of unknown parameters require a greater number of particles. An increased density of particles allows a more effective exploration of the search space and this circumstance implies an enhanced probability of detecting the globally optimal solution (with regard to problems with multiple local minima). This is particularly true in the presence of a fairly large number of inequality constraints, which drastically reduce the number of particles associated with feasible trajectories, thus diminishing the search space exploration capability of the algorithm. The first application considered in this chapter (i.e. the optimal transfer problem between two circular orbits) proves that results of increasing quality can be obtained as the number of particles increases. The remaining space trajectory optimization problems represent very challenging applications for a global optimization algorithm. The number of particles and iterations were set to sufficiently large values, which lead to convergence at the first run. For all of the cases the results attained with the PSO algorithm resemble with great accuracy the numerical solutions previously reported in the scientific literature or known from analysis.

In conclusion, this research demonstrates that the PSO technique indeed represents an efficient, reliable, and accurate method to determine globally optimal space trajectories, despite its intuitiveness and simplicity. However, as alternative heuristic algorithms, the PSO methodology can occasionally encounter difficulties when dealing with constrained problems. In the presence of unsatisfactory results several options are available to a potential user for improving the performance attainable by a PSO algorithm. First, the number of particles and iterations can be increased to achieve more satisfactory results. Second, although the weighting coefficients adopted in this chapter gave satisfactory results for four qualitatively different optimization problems, their fine tuning in the formula for velocity update could yield improved results for alternative applications. Lastly, two versions of the PSO algorithm exist (i.e. the local version and the global version), with their respective advantages and drawbacks. The local version can be employed if the global version is suspected to have yielded a local minimum.

REFERENCES

[1] D. E. Goldberg. *Genetic Algorithms in Search, Optimization, and Machine Learning*, Addison Wesley, Boston, MA, pp. 1—54, 1989.

[2] A.P. Engelbrecht. *Computational Intelligence. An Introduction*, 2nd ed., Wiley, Chichester, UK, pp. 285—411, 555—559, 2007.

[3] R. Eberhart, J. Kennedy, "A New Optimizer Using Particle Swarm Theory", Proceedings of the Sixth International Symposium on Micromachine and Human Science, Nagoya, Japan, 1995.

[4] J. Kennedy, R. Eberhart, "Particle swarm optimization", Proceedings of the IEEE International Conference on Neural Networks, Piscataway, NJ, 1995.

[5] J. Kennedy, R. C. Eberhart. *Swarm Intelligence*, Academic Press, 1st ed., San Diego, CA, pp. 287—324, 2001.

[6] X. Hu, R. Eberhart, "Solving Constrained Nonlinear Optimization Problems with Particle Swarm Optimization", Proceedings of the Sixth World Multiconference on Systemics, Cybernetics and Informatics (SCI 2002), Orlando, FL, 2002.

[7] X. Hu, Y. Shi, R. Eberhart, "Recent Advances in Particle Swarm", Proceedings of the IEEE Congress on Evolutionary Computation (CEC 2004), Portland, OR, 2004.

[8] X. Hu, R. C. Eberhart, Y. Shi, "Engineering Optimization with Particle Swarm", Proceedings of the IEEE Swarm Intelligence Symposium (SIS 2003), Indianapolis, IN, 2003.

[9] R. C. Eberhart, Y. Shi, "Particle Swarm Optimization: Developments, Applications, and Resources", Proceedings of the IEEE Congress on Evolutionary Computation (CEC 2001), Seoul, Republic of Korea, 2001.

[10] R. C. Eberhart, Y. Shi, "Comparing Inertia Weights and Constriction Factors in Particle Swarm Optimization", Proceedings of the IEEE Congress on Evolutionary Computation (CEC 2000), La Jolla, CA, 2000.

[11] A. Carlisle, G. Dozier, "An Off-The-Shelf PSO", Proceedings of the Workshop on Particle Swarm Optimization, Indianapolis, IN, 2001.

[12] M. Clerc, "The Swarm and the Queen: Towards a Deterministic and Adaptive Particle Swarm Optimization", Proceedings of the IEEE Congress on Evolutionary Computation (CEC 1999), Washington, D.C., 1999.

[13] G. Venter, J. Sobieszczanski-Sobieski, "Particle Swarm Optimization", *AIAA Journal*, Vol. 41, No. 8, pp. 1583—1589, 2003.

[14] K. E. Parsopoulos, M. N. Vrahatis, "Particle Swarm Optimization Method for Constrained Optimization Problems", *Intelligent Technologies – Theory and Applications: New Trends in Intelligent Technologies, Frontiers in Artificial Intelligence and Applications series*, Vol. 76, edited by P. Sincak, J. Vascak, V. Kvasnicka, J. Pospichal, pp. 214—220, 2002.

[15] R. Mendes, J. Kennedy, J. Neves, "The Fully Informed Particle Swarm: Simpler, Maybe Better", *IEEE Transactions on Evolutionary Computation*, Vol. 8, No. 3, pp. 204—210, 2004.

[16] S. Kitayama, K. Yamazaki, M. Arakawa, "Adaptive Range Particle Swarm Optimization", paper AIAA 2006-6912, Proceedings of the 11th AIAA/ISSMO Multidisciplinary Analysis and Optimization Conference, Portsmouth, VA, 2006.

[17] N. Higashi, H. Iba, "Particle Swarm Optimization with Gaussian Mutation", Proceedings of the IEEE Swarm Intelligence Symposium (SIS 2003), Indianapolis, IN, 2003.

[18] V. Kalivarapu, E. Winer, "Implementation of Digital Pheromones in Particle Swarm Optimization for Constrained Optimization Problems", paper AIAA 2008-1974, Proceedings of the 49th AIAA/ASME/ASCE /AHS/ASC Structures, Structural Dynamics and Materials Conference, Schaumburg, IL, 2008.

[19] P. J. Angeline, "Evolutionary Optimization Versus Particle Swarm Optimization: Philosophy and Performance Differences", *Evolutionary*

programming VII, *Lecture Notes in Computer Science*, Vol. 1447, Springer, pp. 601—610, 1998.

[20] R. C. Eberhart, Y. Shi, "Comparison Between Genetic Algorithms and Particle Swarm Optimization", *Evolutionary Programming VII*, *Lecture Notes in Computer Science*, Vol. 1447, Springer, 611—616, 1998.

[21] R. Hassan, B. Cohanim, O. de Weck, "A Comparison of Particle Swarm Optimization and the Genetic Algorithm", paper AIAA 2005-1897, Proceedings of the 46th AIAA/ASME/ASCE/AHS/ASC Structures, Structural Dynamics and Materials Conference, Austin, TX, 2005.

[22] P. C. Fourie, A. A. Groenwold, "Particle Swarms in Topology Optimization", *Proceedings of the Fourth World Congress of Structural and Multidisciplinary Optimization*, Liaoning Electronic Press, pp. 1771—1776, 2001.

[23] K. E. Parsopoulos, M. N. Vrahatis, "On the Computation of All Global Minimizers Through Particle Swarm Optimization", *IEEE Transactions on Evolutionary Computation*, Vol. 8, No. 3, pp. 211—224, 2004.

[24] A. R. Cockshott, B. E. Hartman, "Improving the Fermentation Medium for Echinocandin B Production Part II: Particle Swarm Optimization", *Process Biochemistry*, Vol. 36, No. 7, pp. 661—669, 2001.

[25] C. O. Ourique, E. C. Biscaia, J. C. Pinto, "The Use of Particle Swarm Optimization for Dynamical Analysis in Chemical Processes", *Computers and Chemical Engineering*, Vol. 26, No. 12, pp. 1783—1793, 2002.

[26] P. C. Fourie, A. A. Groenwold, "The Particle Swarm Optimization Algorithm in Size and Shape Optimization", *Structural and Multidisciplinary Optimization*, Vol. 23, No. 4, pp. 259—267, 2002.

[27] M. S. Khurana, H. Winarto, A. K. Sinha, "Application of Swarm Approach and Artificial Neural Networks for Airfoil Shape Optimization", paper AIAA 2008-5954, Proceedings of the 12th AIAA/ISSMO Multidisciplinary Analysis and Optimization Conference, Victoria, Canada, 2008.

[28] C. R. Bessette, D. B. Spencer, "Optimal Space Trajectory Design: A Heuristic-Based Approach", *Advances in the Astronautical Sciences*, Univelt Inc., San Diego, CA, Vol. 124, pp. 1611—1628; AAS paper 06-197, 2006.

[29] C. R. Bessette, D. B. Spencer, "Identifying Optimal Interplanetary Trajectories through a Genetic Approach", paper AIAA 2006-6306, AIAA/AAS Astrodynamics Specialist Conference and Exhibit, Keystone, CO, 2006.

[30] M. Vasile, E. Minisci, M. Locatelli, "On Testing Global Optimization Algorithms for Space Trajectory Design", paper AIAA 2008-6277, AIAA/AAS Astrodynamics Specialist Conference and Exhibit, Honolulu, HI, 2008.

[31] K. Zhu, J. Li, H. Baoyin, "Satellite scheduling considering maximum observation coverage time and minimum orbital transfer fuel cost", *Acta Astronautica*, Vol. 66, pp. 220—229, 2010.

[32] K. Zhu, F. Jiang, J. Li, H. Baoyin, "Trajectory Optimization of Multi-Asteroids Exploration with Low Thrust", *Transactions of the Japan Society for Aeronautical and Space Sciences*, Vol. 52, No. 175, pp. 47—54, 2009.

[33] M. Rosa Sentinella, L. Casalino, "Cooperative evolutionary algorithm for space trajectory optimization", *Celestial Mechanics and Dynamical Astronomy*, Vol. 105, No. 1-3, pp. 211—227, 2009.

[34] M. Pontani, B. A. Conway, "Particle Swarm Optimization Applied to Space Trajectories", *Journal of Guidance, Control, and Dynamics*, Vol. 33, No. 5, pp. 1429—1441, 2010.

[35] M. Pontani, B. A. Conway, "Swarming Theory Applied to Space Trajectory Optimization", *Spacecraft Trajectory Optimization*, Cambridge University Press, pp. 263—293, 2010.

[36] M. Pontani, B. A. Conway, "Optimal Space Trajectories via Particle Swarm Technique", *Advances in the Astronautical Sciences*, Vol. 136, pp. 53—72, 2010.

[37] M. Pontani, B. A. Conway, "Particle Swarm Optimization of Low-Thrust Orbital Transfers and Rendezvous", *Advances in the Astronautical Sciences*, Vol. 140, pp. 889—908, 2011.

[38] M. Pontani, B. A. Conway, "Optimal Finite-Thrust Rendezvous Trajectories Found via Particle Swarm Algorithm", *Advances in the Astronautical Sciences*, Vol. 142, pp. 3781—3800, 2012.

[39] M. Pontani, B. A. Conway, "Particle Swarm Optimization Applied to Impulsive Orbital Transfers", *Acta Astronautica*, Vol. 74, pp. 141—155, 2012.

[40] M. Pontani, P. Ghosh, B. A. Conway, "Particle Swarm Optimization of Multiple-Burn Rendezvous Trajectories", *Journal of Guidance, Control, and Dynamics*, Vol. 35, No. 4, 2012, pp. 1192—1207.

[41] M. Pontani, C. Martin, B. A. Conway, "New Numerical Methods for Determining Periodic Orbits in the Circular Restricted Three-Body Problem", Proceedings of the 61st International Astronautical Congress, Prague, Czech Republic, 2010; paper IAC-10-C1.1.2.

[42] Z. Michalewicz, M. Schoenauer, "Evolutionary Algorithms for Constrained Parameter Optimization Problems", *Evolutionary Computation*, Vol. 4, No. 1, pp. 1—32, 1996.

[43] B. Prasad, "A Class of Generalized Variable Penalty Methods for Nonlinear Programming", *Journal of Optimization Theory and Applications*, Vol. 35, No. 2, pp. 159—182, 1981.

[44] R. Fletcher. *Practical methods of optimization*, Wiley, Chichester, UK, pp. 277—325, 1987.

[45] S. Koziel, Z. Michalewicz, "Evolutionary Algorithms, Homorphous Mappings, and Constrained Parameter Optimization", *Evolutionary Computation*, Vol. 7, No. 1, pp. 19—44, 1999.

[46] H. M. Robbins, "Analytical Study of the Impulsive Approximation", *AIAA Journal*, Vol. 4, No. 8, pp. 1417—1423, 1966.

[47] C. Zee, "Effects of Finite-thrusting Time in Orbit Maneuvers", *AIAA Journal*, Vol. 1, No. 1, pp. 60—64, 1963.

[48] J.-P. Marec. *Optimal Space Trajectories*, Elsevier, Amsterdam, the Netherlands, pp. 21—27, 1979.

[49] G. A. Hazelrigg, "Globally Optimal Impulsive Transfers via Green's Theorem", *Journal of Guidance, Control, and Dynamics*, Vol. 7, No. 4, pp. 462—470, 1984.

[50] R. H. Battin. *An Introduction to the Mathematics and Methods of Astrodynamics*, AIAA Education Series, AIAA, New York, NY, pp. 529—530, 1987.

[51] J. E. Prussing, "Simple Proof of the Global Optimality of the Hohmann Transfer", *Journal of Guidance, Control, and Dynamics*, Vol. 15, No. 4, pp. 1037—1038, 1992.

[52] J. E. Prussing, B. A. Conway. *Orbital Mechanics*, Oxford University Press, New York, NY, pp. 62—80, 103—112, 1993.

[53] M. Pontani, "Simple Method to Determine Globally Optimal Orbital Transfers", *Journal of Guidance, Control, and Dynamics*, Vol. 32, No. 3, pp. 899—914, 2009.

[54] D. Stratemeier, "Optimum Two-Impulse Orbital Transfer Solved Using Evolutionary Programming", paper AIAA 2002-4908, Proceedings of the AIAA/AAS Astrodynamics Specialist Conference and Exhibit, Monterey, CA, 2002.

[55] J. E. Prussing, "Optimal Four-Impulse Fixed-Time Rendezvous in the Vicinity of a Circular Orbit", *AIAA Journal*, Vol. 7, No. 5, pp. 928—935, 1969.

[56] J. E. Prussing, "Optimal Two- and Three-Impulse Fixed-Time Rendezvous in the Vicinity of a Circular Orbit", *AIAA Journal*, Vol. 8, No. 7, pp. 1221—1228, 1970.

[57] J. E. Prussing, J.-H. Chiu, "Optimal Multiple-Impulse Time-Fixed Rendezvous Between Circular Orbits", *Journal of Guidance, Control, and Dynamics*, Vol. 9, No. 1, pp. 17—22, 1986.

[58] G. Colasurdo, D. Pastrone, "Indirect optimization method for impulsive transfer", paper AIAA 94-3762, Proceedings of the AIAA/AAS Astrodynamics Conference, Scottsdale, AZ, 1994.

[59] A. B. Mickelwait, R. C. Booton, "Analytical and numerical studies of three-dimensional trajectories to the Moon", *Journal of the Aerospace Sciences*, Vol. 27, No. 8, pp. 561—573, 1960.

[60] V. C. Clarke, "Design of lunar and interplanetary ascent trajectories", *AIAA Journal*, Vol. 5, No. 7, pp. 1559—1567, 1963.

[61] C. S. Dallas, "Moon-to-Earth trajectories", *Celestial Mechanics and Dynamical Astronomy*, Vol. 14, pp. 391—438, 1964.

[62] N. H. Shaikh, "A new perturbation method for computing Earth-Moon trajectories", *Astronautica Acta*, Vol. 12, No. 3, pp. 207—211, 1966.

[63] V. Szebehely, *Theory of Orbits in the Restricted Problem of Three Bodies*, Academic Press, London, 1967.

[64] M. Pontani, P. Teofilatto, "Low-Energy Earth-Moon Transfers Involving Manifolds", Proceedings of the 63rd International Astronautical Congress, Naples, Italy, 2013; paper IAC-12-C1.7.1.

[65] M. Giancotti, M. Pontani, P. Teofilatto, "Earth-Moon Transfers Involving Periodic Orbits and Invariant Manifolds through Isomorphic Mapping", Proceedings of the 13th International Space Conference of Pacific-basin Societies, Kyoto, Japan, 2012.

[66] M. W. Lo, B. G. Williams, W. E. Bollman, D. S. Han, Y. S. Hahn, J. L. Bell., E. Hirst, R. Corwin, R. Hong, K. Howell, B. Barden, R. Wilson, "Genesis mission design", *Journal of the Astronautical Sciences*, Vol. 49, No. 1, pp. 169–184, 2011.

[67] C. A. Kluever, B. L. Pierson, "Optimal Earth-Moon trajectories using nuclear electric propulsion", *Journal of Guidance, Control, and Dynamics*, Vol. 20, No. 2, pp. 239—245, 1997.

[68] C. A. Kluever, "Optimal Earth-Moon trajectories using combined chemical-electric propulsion", *Journal of Guidance, Control, and Dynamics*, Vol. 20, No. 2, pp. 253—258, 1997.

[69] A. Miele, S. Mancuso, "Optimal Trajectories for Earth-Moon-Earth Flight", *Acta Astronautica*, Vol. 49, No. 2, pp. 59—71, 2001.

In: Global Optimization
Editor: Angelika Michalski

ISBN: 978-1-62417-796-5

Chapter 3

Float-Encoded Genetic Algorithm Used for the Inversion Processing of Well-Logging Data

***Norbert P. Szabó*[1]* *and Mihály Dobróka*[1,2]**
[1]University of Miskolc, Department of Geophysics, Hungary
[2]MTA-ME Research Group for Earth Science and Engineering, Hungary

Abstract

A Float-Encoded Genetic Algorithm is presented in this chapter for solving the well-logging inverse problem. The aim of the global inversion of well-logging data is to provide a robust and reliable estimate of petrophysical properties, such as porosity, water saturation, shale volume and mineral content, associated with geological structures. There are two possible ways to solve the interpretation problem. The first is a conventional inversion scheme, separately estimating the unknowns to different depths. In the forward modeling phase of the local inversion procedure the theoretical well-logging data set is calculated by using locally defined probe response functions, which are then fit to real data in order to estimate model parameters to one depth only. This procedure leads to a marginally over-determined inverse problem, which results in relatively poor parameter estimates. A further disadvantage of the latter technique is that some crucial quantities, such as the thickness of layered geological formations, cannot be extracted by inversion because they do

* Corresponding author's e-mail address: norbert.szabo.phd@gmail.com

not appear explicitly in local response equations. A new inversion methodology introduced by the authors gives much more freedom in choosing the inversion parameters. The so-called interval inversion method inverts all data measured from a greater depth interval in a joint inversion process. By a series expansion-based discretization of the petrophysical model, a highly over-determined inverse problem can be formulated, enabling an estimation of the petrophysical parameters including new unknowns, such as zone parameters and layer thicknesses, more accurately compared to local inversion methods. The authors give further references for several applications of the global inversion method. In this chapter, one synthetic and two field examples are presented to demonstrate the application of the Genetic Algorithm-based inversion method. It is shown that the combination of the new inversion strategy and global optimization tools forms a highly effective and adaptive algorithm for earth scientists who are interested in a more reliable calculation of the reserves of hydrocarbons and other mineral resources.

Keywords: Genetic Algorithm, well-logging, petrophysical model, interval inversion

1. INTRODUCTION

Well-logging methods are extensively used in mineral exploration for collecting high resolution *in-situ* information associated with subsurface structures. Surveying methods are based on different physical principles, which result in well logs of several measured quantities (Serra, 1984; Asquith and Krygowski, 2004). The main object of a log analysis is the lithologic separation of the succession of strata and the estimation of several geometrical (layer-thickness, formation dip, lateral changes of layer boundaries) and petrophysical properties of geologic formations (porosity, water saturation, mineral content, permeability). The most advanced data processing tools are based on geophysical inversion methods. Both the principles and computer implementation of the well-logging inversion procedure are detailed by Mayer and Sibbit (1980), Alberty and Hashmy (1984) and Ball et al. (1987).

The interpretation of well-logging data sets is traditionally solved by linearized inversion methods (Menke, 1984; Tarantola 2005). Linear optimization methods work properly and quickly, provided that we have satisfactory *a priori* information about the petrophysical model. Well-logging measurements are usually supported by laboratory data measured on core samples and independent results of geophysical measurements made on the surface (Telford et al., 1990). Although there is an abundance of information available, the data noise and inherent ambiguity of the inverse problem often cause the failure of the linear optimization procedure as it is trapped in a local minimum solution. The problem can be effectively avoided by using global optimization methods that search for the absolute extreme of the multivariate objective function. Currently, the most preferred global optimization procedures are the very fast algorithms of the Simulated Annealing method (Metropolis et al., 1953; Sen and Stoffa, 1997) and the Float-Encoded Genetic Algorithm (Michalewicz, 1992).

The Genetic Algorithm (GA) was proposed by John Holland (1975) for solving optimization problems by using the analogy of natural selection in biology. The procedure improves a population of random models in an iteration process. Each individual of a given generation has a fitness value, which represents its survival capability. In the geophysical inverse problem the fitness function is connected to the distance between the observed data and theoretical data calculated on the actual petrophysical model. The most probable petrophysical model can be estimated by maximizing the fitness function (i.e., minimization of data misfits). During the genetic process the fittest individuals reproduce more successfully in the subsequent generations than those who have less fitness. In the last generation, the individual with the maximum fitness corresponds to the solution of the optimization problem. In the classical GA procedure the model parameters are encoded using a binary coding scheme, setting a limit to the resolution of the petrophysical model and the accuracy of the estimation results. The Float-Encoded GA computes with model parameters as real numbers, which is faster than binary coding, because it does not use coding-decoding phases and achieves higher resolution of the model space (Houck et al., 1996).

Conventional well-logging inversion methods process the data set of a certain measuring point along the borehole to determine the petrophysical model parameters only to that point. This local inversion technique represents a narrow type of the over-determined inverse problem because the number of probe types is slightly more than that of the unknowns. This inversion strategy leads to a set of separate inversion runs in adjacent depth points for the logging interval. The marginal over-determination sets a limit to the accuracy and reliability of the estimation, and does not support the determination of too many petrophysical parameters. A novel inversion methodology for solving the above problems was developed by the Department of Geophysics, University of Miskolc. The so-called interval inversion method describes the petrophysical model parameters in the form of a series expansion. By this formulation the validity of probe response functions used in local forward modeling is extended to a greater depth interval. Thus, well-logging data of an optional depth interval can be inverted jointly to give an estimate for the model parameters to the same interval. The joint inversion procedure has at least an order of magnitude greater over-determination (data to unknowns) ratio compared to local inversion, which results in a significant improvement in the quality of interpretation.

The interval inversion method has been applied to different interpretation problems related to hydrocarbon exploration (Dobróka and Szabó, 2001; Szabó et al., 2003; Szabó, 2004; Dobróka and Szabó, 2005; Dobróka et al., 2005; Dobróka et al., 2007; Dobróka et al., 2009; Dobróka and Szabó, 2011; Dobróka et al. 2012; Dobróka and Szabó, 2012). Because of its high performance and adaptability we chose the GA for solving the interval inversion problem, which has proven to give more accurate estimates than derivative-based linear inversion procedures and results in a practically independent solution of the initial model. This chapter presents the GA-based interval inversion algorithm applied to a synthetic inversion experiment and two field cases.

2. The Inversion Methodology

2.1. Forward Modeling

In the forward modeling phase of the well-logging inversion procedure a theoretical data set is predicted from the petrophysical model. Consider the column vector of the model parameters

$$\mathbf{m} = \left[\Phi, S_{x0}, S_w, V_{sh}, V_{ma,l}\right]^T, \tag{2.1}$$

where Φ denotes the porosity of the geological formation, S_{x0} is the water saturation in the nearest zone around the borehole where pore space is invaded by drilling mud, S_w is the water saturation in the undisturbed zone filled with the original pore fluid (water and/or hydrocarbon), V_{sh} is the shale volume, and $V_{ma,l}$ is the volume of the l-th mineral component forming the rock matrix (superscript T denotes matrix transpose). Each parameter in (2.1) is a dimensionless quantity given in a fraction or per cent.

$$\mathbf{d}^{(meas.)} = \left[SP, GR, \Phi_N, \rho_b, \Delta t, R_s, R_d\right]^T, \tag{2.2}$$

where *SP (mV)* denotes the spontaneous potential (lithology log), *GR (API)* is the natural γ-ray intensity (lithology log), Φ_N *(%)* is the neutron porosity (porosity log), Δt *(μs/m)* is the acoustic traveltime (porosity log), ρ_b *(g/cm³)* is the formation density (porosity log), and R_s and R_d *(ohm-m)* are electric resistivities measured with shallow and deep penetration tools (saturation logs), respectively. The relationship between the model (2.1) and data (2.2) corrected for borehole environmental effects is called the probe response function (**g**). The data vector approximated by the calculation is:

$$\mathbf{d}^{(calc.)} = \mathbf{g}(\mathbf{m}). \tag{2.3}$$

In practice, equation (2.3) is given by a set of empirical equations including the petrophysical parameters and some local specific constants chosen according to the geological setting.

The following simplified response equations based on Poupon and Leveaux (1971) and Alberty and Hashmy (1984) can be used to calculate data in the forward modeling procedure in case of shaly-sand hydrocarbon reservoirs:

$$SP^{(calc.)} = SP_{sd} + V_{sh}\left(SP_{sh} - SP_{sd}\right), \tag{2.4}$$

$$GR^{(calc.)} = \Phi\left[GR_{mf} S_{x0} + GR_{hc}\left(1 - S_{x0}\right)\right] + V_{sh} GR_{sh} + V_{sd} GR_{sd}, \tag{2.5}$$

$$\Phi_N^{(calc.)} = \Phi\left[\Phi_{N,mf} S_{x0} + \Phi_{N,hc}\left(1 - S_{x0}\right)\right] + V_{sh}\Phi_{N,sh} + V_{sd}\Phi_{N,sd}, \tag{2.6}$$

$$\rho_b^{(calc.)} = \Phi\left[\rho_{b,mf} S_{x0} + \rho_{hc}\left(1 - S_{x0}\right)\right] + V_{sh}\rho_{b,sh} + V_{sd}\rho_{b,sd}, \tag{2.7}$$

$$\Delta t^{(calc.)} = \Phi\left[\Delta t_{mf} S_{x0} + \Delta t_{hc}\left(1 - S_{x0}\right)\right] + V_{sh}\Delta t_{sh} + V_{sd}\Delta t_{sd}, \tag{2.8}$$

$$\frac{1}{\sqrt{R_s^{(calc.)}}} = \left[\frac{V_{sh}^{(1-0.5V_{sh})}}{\sqrt{R_{sh}}} + \frac{\left(\sqrt{\Phi}\right)^m}{\sqrt{aR_{mf}}}\right]\left(\sqrt{S_{x0}}\right)^n, \tag{2.9}$$

$$\frac{1}{\sqrt{R_d^{(calc.)}}}=\left[\frac{V_{sh}^{(1-0.5V_{sh})}}{\sqrt{R_{sh}}}+\frac{\left(\sqrt{\Phi}\right)^m}{\sqrt{aR_w}}\right]\left(\sqrt{S_w}\right)^n, \tag{2.10}$$

where V_{sd} denotes the volume of quartz matrix (sand), m is the cementation exponent, n is the saturation exponent, and a is the tortuosity factor. Subscripts appearing in equations (2.4)-(2.10) denote physical properties of mud filtrate (*mf*), hydrocarbon (*hc*), shale (*sh*) and sand (*sd*), which can be found in literature or *a priori* information of the measurement area. Some response function constants can be determined by the interval inversion procedure (Dobróka and Szabó, 2011), but this chapter does not deal with that problem. Layer-thicknesses are not included explicitly in local response equations; thus, they can only be determined from a distinct procedure (e.g., by manual log analysis). Obviously, the above set of equations is non-linear with respect to the petrophysical model; thus, it is advantageous to solve the inverse problem by means of global optimization. Moreover, the simple structure of equations (2.4)-(2.10) enables the calculation of data in a relatively fast procedure, which is highly favorable for using GA.

2.2. The Inverse Problem

2.2.1. Local Inversion Methods

The non-linear well-logging inverse problem is conventionally solved by local inversion methods that separately give an estimate for the petrophysical model at adjacent depth points. The calculated data of the j-th well log in a given depth from equation (2.3) is

$$d_j^{(calc.)}=g_j\left(m_1,\ldots,m_M\right), \tag{2.11}$$

where M is the number of model parameters in equation (2.1). Response function constants are not indicated in equation (2.11). They are defined for a longer depth interval, describing rock formations of the same origin and pore-fluid properties, thereby named zone parameters. Theoretical data can be calculated by equations (2.4)-(2.10). The difference between the observations and predictions is minimized in an iterative procedure. The objective function of the optimization problem is based on the Euclidean norm of the overall data misfit (Mayer and Sibbit, 1980). Since the uncertainty of data is different, it is advantageous to use weighting in the data space. The standard deviation of data, depending on the probe type and borehole conditions, can be found in the literature. The linear optimization approach is popular in industrial applications because it gives a quick solution to the inverse problem. On the other hand, in the case of the Gaussian (linear) least squares solution, the data and model variances can be related to each other; thus, the estimation errors of model parameters can be quantified (Menke, 1984). The quality check of inversion estimates is an important issue in well-log analysis.

2.2.2. Interval Inversion Method

In local inverse problems the amount of data is barely more than that of the unknowns, which results in a relatively noise sensitive inversion procedure. To preserve the over-determination of the inverse problem the zone parameters are required to be fixed in the procedure. The accuracy of the solution highly depends on the noise level of data and the initial guess of the petrophysical model. To improve the over-determination ratio we extend the validity of function (2.11) to a greater depth interval,

$$d_j^{(calc.)}(z) = g_j\left(m_1(z), m_2(z), \ldots, m_M(z)\right), \tag{2.12}$$

which can be used to calculate the j-th data at an arbitrary depth. In equation (2.12) model parameters are represented by continuous functions that have to be properly discretized. (In the general case zone parameters are also the function of depth.) The interval inversion method is based on the series expansion of petrophysical parameters (Dobróka, 1993):

$$m_i(z) = \sum_{q=1}^{Q_i} B_q^{(i)} \Psi_q(z), \tag{2.13}$$

where m_i denotes the i-th model parameter, B_q is the q-th expansion coefficient and Ψ_q is the q-th basis function (Q_i is the requisite number of expansion coefficients describing the i-th model parameter). Basis functions are assumed to be known quantities, which can be arbitrarily chosen. In this chapter, we present the simplest case when layer-wise homogeneous layers build up a sequence. The following combination of unit step functions divides the processed depth interval into homogeneous segments

$$\Psi_q(z) = u(z - Z_{q-1}) - u(z - Z_q), \tag{2.14}$$

where Z_{q-1} and Z_q are the upper and lower depth coordinates of the q-th layer, respectively. Since $\Psi_q(z)$ is always zero, except in the q-th layer (where $\Psi_q(z)=1$), each petrophysical parameter can be described by one series expansion coefficient $m_i(z_{q-1} \leq z \leq z_q) = B_q^{(i)}$. The model parameter vector of the inverse problem is

$$\mathbf{m} = \left[B_1^{(1)}, \ldots, B_Q^{(1)}, \ldots\ldots, B_1^{(M)}, \ldots, B_Q^{(M)}, Z_1, \ldots, Z_{Q-1}\right]^T, \tag{2.15}$$

where Q is the number of layers. The advantage of the interval-wise homogeneous model approximation is that an amount of data significantly higher than that of the unknowns can be integrated into a greatly over-determined inversion procedure. The latter also provides the opportunity to treat layer-boundary coordinates as inversion unknowns, which is not possible to extract by local inversion. The inversion method developed for the automatic estimation of layer-thicknesses was introduced by Dobróka and Szabó (2012).

Well-logging data have different magnitudes and measurement units. We introduce a weighted objective function based on the least squares criterion for the interval inversion process:

$$E(\mathbf{m}) = \sum_{p=1}^{P} \sum_{j=1}^{N} \left(\frac{d_{pj}^{(meas.)} - d_{pj}^{(calc.)}}{d_{pj}^{(meas.)}} \right)^2 = \min, \tag{2.16}$$

where $d_{pj}^{(meas.)}$ and $d_{pj}^{(calc.)}$ are the j-th measured and calculated data in the p-th depth, respectively (P is the number of measuring points in the processed interval and N is the number of logging instruments). The optimal values of the series expansion coefficients in equation (2.15) are estimated by the optimization of function E, which are then substituted into equation (2.13) to produce well logs of petrophysical parameters.

3. GENETIC ALGORITHM OPTIMIZATION

The Genetic Algorithm search is based on an analogy to the process of natural selection of living populations (Holland, 1975). For optimization problems, the model can be considered as an individual of an artificial population, where the fitness characterizes the goodness of the solution. A real-valued GA approach, based on Michalewicz (1992), is applied for seeking the absolute minimum of the objective function of the well-logging inverse problem. The procedure operates with model parameters encoded as floating-point numbers that provide the highest precision estimation and best CPU time performance of all GAs.

The survival capability of an individual is characterized by a fitness value, which shows its chance of reproduction into the next generation. The fitness function is inversely connected to objective function (2.16):

$$F(\mathbf{m}) = \frac{1}{\sum_{p=1}^{P}\sum_{j=1}^{N}\left(\frac{d_{pj}^{(meas.)} - d_{pj}^{(calc.)}}{d_{pj}^{(meas.)}}\right)^2 + \varepsilon^2} = \max\,, \tag{3.1}$$

where ε^2 is a small positive number used for setting a limit of model fitness. Function (3.1) is inversely proportional to the Euclidean norm of data misfits. In the case of non-Gaussian data sets, including outliers, the L_1-norm based fitness function gives the more robust solution,

$$F(\mathbf{m}) = \frac{1}{\sum_{p=1}^{P}\sum_{j=1}^{N}\left|\frac{d_{pj}^{(meas.)} - d_{pj}^{(calc.)}}{d_{pj}^{(meas.)}}\right| + \varepsilon^2} = \max\,. \tag{3.2}$$

The iteration procedure is convergent if the average fitness of the individuals increases in successive generations. This is assured by the appropriate choice of genetic operations and their control parameters.

After setting up an initial population of petrophysical models, a random search is launched for finding the optimal solution. In the procedure, three types of genetic operations are performed on the current population: selection, crossover and mutation. At first, the fittest individuals are selected from the population. We apply a normalized geometric ranking selection that requires the individuals to be sorted according to their fitness values. The rank of the best individual is 1 and that of the worst is S, which is the size of the population. The probability of selecting the x-th individual is

$$P_x = \frac{p_b}{1-(1-p_b)^S}(1-p_b)^{r_x-1}, \tag{3.3}$$

where r_x is the rank of the x-th individual, and p_b is the probability of selecting the best individual.

The latter quantity is a control parameter which has to be set in advance. The x-th individual is selected and copied into the new population when the following condition fulfills

$$C_{x-1} < U \leq C_x, \tag{3.4}$$

where U is a uniform random number between 0 and 1, and

$$C_x = \sum_{y=1}^{x} P_y \tag{3.5}$$

is the cumulative probability of the population.

In the next step, a pair of individuals is chosen from the previously selected population and a partial information exchange is made between them. The simple crossover operator recombines the original individuals to create new ones:

$$\begin{aligned} \mathbf{m}^{(1,new)} &= \begin{cases} m_i^{(1,old)} & \text{if } i<k \\ m_i^{(2,old)} & \text{otherwise} \end{cases} \\ \mathbf{m}^{(2,new)} &= \begin{cases} m_i^{(2,old)} & \text{if } i<k \\ m_i^{(1,old)} & \text{otherwise} \end{cases} \end{aligned}, \tag{3.6}$$

where k is the position of the crossing point, and $m_i^{(old)}$ and $m_i^{(new)}$ denote the i-th model parameter before and after crossover, respectively.

The third genetic operator is mutation, which selects an individual from the population and changes one (or more) of its model parameters to a random number. A simple mutation operator modifies the original model as follows:

$$\mathbf{m}^{(new)} = \begin{cases} u & \text{if } i = k \\ m_i^{(old)} & \text{otherwise} \end{cases}, \tag{3.7}$$

where u is a uniform random number generated from the allowed range of the k-th model parameter. More types of real genetic operations are detailed in Houck et al. (1996).

A quality check of individuals during the inversion procedure is also required. We use two quantities for the characterization of a misfit. The relative data distance based on equation (2.16) is the measure of fit between measured and calculated data

$$D_d = \sqrt{\frac{1}{P \cdot N} \sum_{p=1}^{P} \sum_{j=1}^{N} \left(\frac{d_{pj}^{(meas.)} - d_{pj}^{(calc.)}}{d_{pj}^{(meas.)}} \right)^2} . \tag{3.8}$$

The relative model distance characterizes the goodness of the estimated model in cases of synthetic inversion experiments:

$$D_m = \sqrt{\frac{1}{QM} \sum_{q=1}^{Q} \sum_{i=1}^{M} \left(\frac{m_{qi}^{(ex.)} - m_{qi}^{(est.)}}{m_{qi}^{(ex.)}} \right)^2} , \tag{3.9}$$

where $m_{qi}^{(est.)}$ and $m_{qi}^{(ex.)}$ are the i-th estimated and exactly known model parameters in the q-th interval, respectively (Q is the number of homogeneous layers along the logging interval and M is the number of model parameters). Quantities (3.8) and (3.9) are usually multiplied by 100 to measure the misfit percentage.

Figure 1 illustrates the workflow of the GA-based inversion procedure. After collecting all measurement data and *a priori* information we set the searching intervals of the parameters of the petrophysical model. Normally, an initial population of 30-100 models is randomly generated from the search space. In the forward modeling phase of the inversion procedure, a synthetic well-logging data set is calculated for each model. The data sets are then compared with real measurements. As long as the misfit between the observation and prediction is too high, the model population is improved by using the three genetic operations. The development of convergence is highly dependent on the proper setting of control parameters of genetic operators. The searching process stops when a termination criterion is met. This can be defined by the maximum number of iterations (generations) or by a specified threshold in the distance between measured and calculated data. In the last iteration step we accept the fittest individual of the generation as the optimal solution.

The above presented GA-based inversion scheme forms an effective algorithm, but it requires more CPU time than Simulated Annealing or linear optimization methods. This is realated to the processor speed, which can be improved by using stronger workstations. Another disadvantage of GA is the possibility of a lack of quality control from one inversion program run.

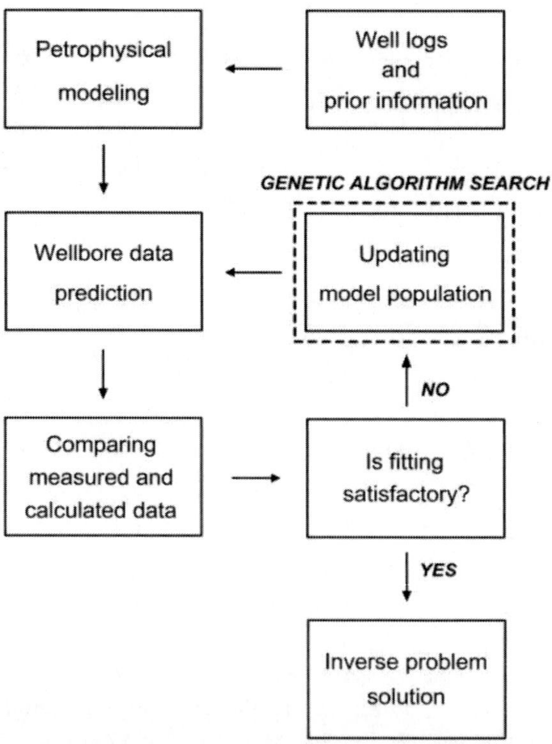

Figure 1. The scheme of GA-based inversion of well-logging data.

Linear optimization methods are based on the calculation of numerical derivatives of data with respect to model parameters, which can be applied to derive estimation errors of model parameters (Menke, 1984). These quantities cannot be extracted by a single GA process, but by using the subsequent combination of GA and a proper linear optimization method both the speed-up and accuracy quantification can be accomplished (Dobróka and Szabó, 2005).

4. Inversion Experiment Using Synthetic Data

The aim of the inversion of synthetic data is to test the performance of the inversion procedure. The experiment assumes that the petrophysical model is exactly known. Synthetic well logs are calculated on the model, which are then contaminated with some amount of noise to simulate real measurements. This noisy data provides input for the GA-based inversion procedure. It can be studied how accurately and reliably the inversion parameters have returned back to the optimum, and how successfully the known model has been recovered. We used our own developed program code written in MATLAB to perform the GA inversion tests.

We modeled four shaly-sandy water and hydrocarbon-bearing (HC) formations with 6m, 2m, 8m and 4m thicknesses. The parameters of the layer-wise homogeneous model are the unknowns of the inverse problem (Table 1).

Table 1. The petrophysical model for GA inversion of synthetic data

Layer	Z(m)	Φ	S_{x0}	S_w	V_{sh}	V_{sd}	Lithology
1	6.0	0.2	0.8	0.4	0.3	0.5	Shaly sand HC reservoir
2	8.0	0.1	1.0	1.0	0.8	0.1	Shale with small sand
3	16.0	0.3	0.8	0.3	0.1	0.6	Sand HC reservoir with small shale
4	20.0	0.1	1.0	1.0	0.6	0.3	Sandy shale

The synthetic data set was generated by using equations (2.4)-(2.10), which was contaminated by 5% Gaussian distributed noise. The depth interval between adjacent observation points was 0.1 m. The noisy synthetic well logs are in Figure 2. In the case of local inversion we had 5 unknowns and 7 pieces of data in each measuring point, whereas the interval inversion treated 24 unknowns and 1,400 pieces of data in the 20m long interval. It was expected that the significant increase of the data-to-unknowns (over-determination) ratio (from 7/5=1.4 to 1400/24=58.3) would make an improvement in the accuracy of inversion estimates.

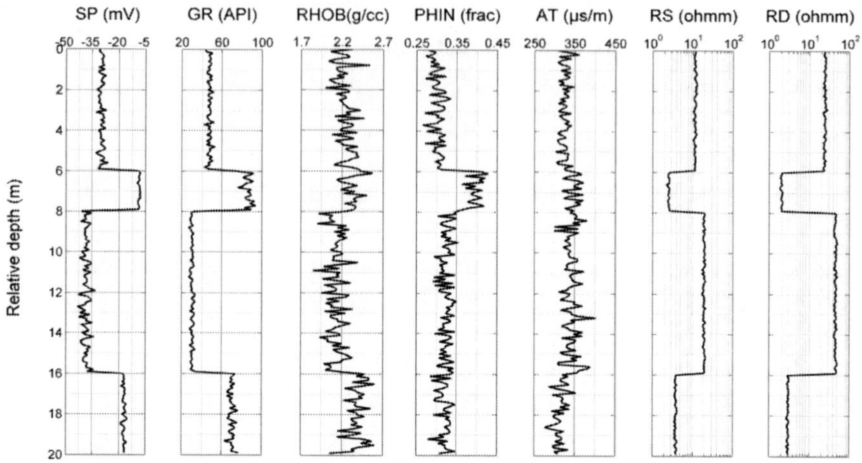

Figure 2. The synthetic well-logging data set contaminated with 5% Gaussian noise serving as input for the GA inversion experiment. Denotations: spontaneous potential (SP), natural γ-ray intensity (GR), density (RHOB), neutron porosity (PHIN), acoustic traveltime (AT), shallow resistivity (RS), deep resistivity (RD).

At first the well logs were processed by the GA-based local inversion method. The search interval of each model parameter was defined between 0 and 1. In each depth, the maximum number of generations was set to 3,500 (iteration steps). The initial population was set to 20 individuals. Real-valued genetic operators (3.3), (3.6), (3.7) were used to improve the petrophysical model population. The control parameters of genetic operators were the probability of selecting the best individual (p_b=0.08) and crossover retry (50) used when an individual is out of bounds, defined by the material balance equation and mutation probability (p_m=0.05). An elitism-based reproduction was applied as the individual with the maximum fitness value was automatically copied into the next generation. As an example, the relative data distances calculated by equation (3.8) in the first measuring point are listed in Table 2. In the first iteration step the average value of data distances was D_d=167.3% beside the standard deviation σ=360.1. Individual no. 12 represents a physically non-realistic model, which does not satisfy the material balance equation. These models are selected out of the population at once. In the 150th generation the average data distance and standard deviation decreased to D_d=23.7% and σ=26.3, respectively. As the process converged towards the optimum, more and more individuals of high fitness values appeared in the population. The average data and model distance of the optimal model were D_d=5.6% and D_m=2.3%, respectively.

Table 2. The selected model generations in the local GA inversion procedure

Ordinal number of individual	Data distance (1st generation)	Data distance (150th generation)
1	96.4 %	10.2 %
2	29.5 %	10.2 %
3	30.3 %	111.0 %
4	96.1 %	17.6 %
5	30.3 %	6.0 %
6	98.9 %	13.6 %
7	83.8 %	7.6 %
8	89.7 %	64.5 %
9	156.7 %	7.8 %
10	122.8 %	7.8 %
11	112.4 %	46.1 %
12	1,690.2 %	46.1 %
13	156.3 %	38.4 %
14	96.0 %	16.4 %
15	75.0 %	16.4 %
16	71.7 %	17.7 %
17	80.9 %	6.0 %
18	68.6 %	6.0 %
19	80.9 %	18.3 %
20	80.9 %	6.0 %

The local inversion results are in Figure 3/a-b, which are as noisy as the input data. From the point of view of the hydrocarbon reserve calculation, further important petrophysical quantities can be derived from inversion estimates. The irreducible ($S_{hc,irr}$) and movable hydrocarbon saturations ($S_{hc,m}$) are:

$$S_{hc,irr} = 1 - S_{x0}, \quad (4.1)$$

$$S_{hc,m} = S_{hc} - S_{hc,irr} = (1 - S_w) - (1 - S_{x0}) = S_{x0} - S_w, \tag{4.2}$$

which are also uncertain at low over-determination ratios. In Figure 3 the above quantities are also plotted. The hydrocarbon saturation is 0% when the pores are completely filled with water (S_{x0}=S_w=1).

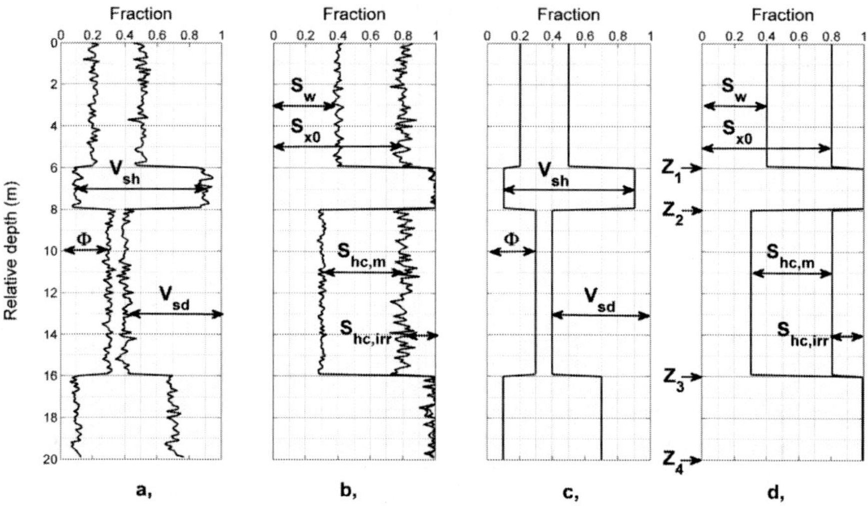

Figure 3. The GA inversion outputs for a layer-wise homogeneous petrophysical model: a-b, local inversion results, c-d, interval inversion results. Denotations: porosity (Φ), shale volume (V_{sh}), sand volume (V_{sd}), water saturation in the undisturbed zone (S_w), water saturation in the flushed zone (S_{x0}), movable hydrocarbon saturation ($S_{hc,m}$), irreducible hydrocarbon saturation ($S_{hc,irr}$), layer-boundary coordinate (Z).

In the next step, the interval inversion method was tested. GA was used to maximize function (3.1). We set the maximal number of generations to 30,000 and fixed the size of the model population to 20 individuals. Genetic operators (3.3), (3.6) and (3.7) were implemented with the same values of control parameters as in the local inversion procedure. Not only were the petrophysical parameters allowed to vary in the procedure, but layer-thicknesses were allowed to vary, too. The search space of layer-thicknesses was specified over the domain of real numbers between 0.1 and 10 (m).

The samples of petrophysical variables were chosen from the interval of 0 and 1. The gradual decrease of data and model distances can be seen in Figure 4. Around iteration step 90, the model distance curve calculated by equation (3.9) shows an escape from a local minimum region.

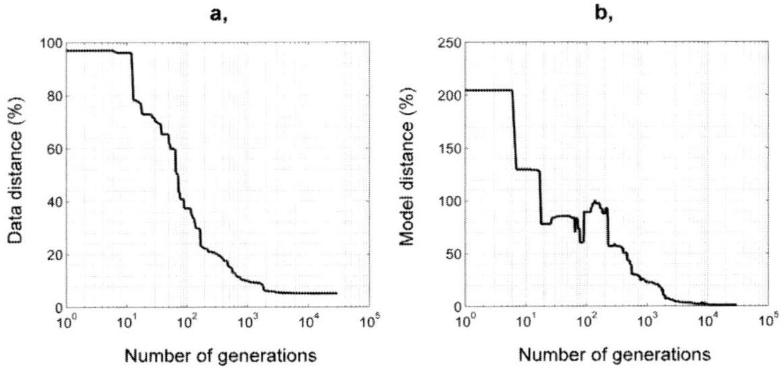

Figure 4. The convergence plots of the GA-based interval inversion procedure: a, data distance of the best individual vs. iteration step, b, model distance of the best individual vs. iteration step.

This event is connected to the finding of optimal layer boundary coordinates. After the 300th generation, only the petrophysical parameters were refined, which proved the stability of the interval inversion procedure.

The optimum was found in the last generation, where the average data distance stabilized around the noise level (D_d=5.6%) and at an average model distance D_m=1.1%. The interval inversion results, including reservoir parameters and formation boundary-coordinates, are shown in Figure 3/c-d. The estimated formation thicknesses were accurate to the third decimal place. Petrophysical model parameters were also obtained close to their exact values (Table 1). The target model represents a real petroleum geological situation, to which a similar field example can be found in Section 5.2. It can be mentioned that the over-determination ratio of the interval inversion problem decreases from 70 to 58.3 by treating formation thicknesses as an unknown model parameter during inversion, but it does not cause a significant quality loss in the overall estimation. This case is still more favorable than the marginally over-determined local inverse problem. By comparing local and interval inversion results, it can be stated that the latter gives more accurate estimates of movable and irreducible hydrocarbon reserves in the case of near-homogeneous reservoirs.

5. Field Cases

The test sites for inversion method development were chosen in the Great Hungarian Plain, Hungary. The study areas are part of the Pannonian Basin

Province of Central Europe, a highly explored area, where many petroleum systems have been discovered. In the Pannonian Basin a thick Tertiary sedimentary sequence overlays the Mesozoic, Paleozoic and Precambrian basement. Most hydrocarbon reservoirs in the area are high or medium-porosity sandstones interbedded with clay, silt, marl and other layers. The oil- and gas-bearing formations situated mostly between 1,000-3,000m in depth represent a wide variety of structural, stratigraphical and combined traps in the province (Dolton 2006). Above the Tertiary basin-fill sequence, there are various Pannonian aged postrift sediments including thermal-water resources. The succeeding Quaternary sediments are characterized by a high variety of paludal, fluvial and delta-plain deposits including thick gravel and sand deposits saturated with fresh water. In this chapter, two case studies about the interval inversion processing of well-logging data originating from shallow water prospecting and deep hydrocarbon wells are presented.

5.1. Application to Water Prospecting

The first example shows the inversion processing of well-logging data collected from a water prospecting borehole (Well-1). As the Earth's crust is relatively thin compared to surrounding mountainous areas, there is a high heat flow in the region. The high amount of thermal water reserves is the source of a very significant and perspective geothermal energy. On the other hand, the same water is pumped into injection wells to improve oil recovery from the underlying hydrocarbon reservoirs. We processed the data of a short 19.4-meter interval from the thermal water zone. The well logs involved in the interval inversion procedure are in Figure 5. Measurement types can be found in equation (2.2). (The microlaterolog - *RMLL* - is a focused type of resistivity measurement with a very shallow radial depth of penetration.)

We assumed four beds by the GR log. Figure 6/a shows the result of manual separation of sedimentary layers. The lithology is also included in the figure, which was confirmed by prior geological knowledge of the area. We applied the interval inversion method to determine porosity, shale content, sand volume and layer boundaries. The GA search was performed with a population of 20 individuals and 12,000 generations. Genetic operators (3.3), (3.6), (3.7) were applied by the following control parameters: probability of selecting the best individual (p_b=0.08), crossover retry (200) and mutation probability (p_m=0.05). The decrease of data distance during the inversion procedure can be seen in Figure 7.

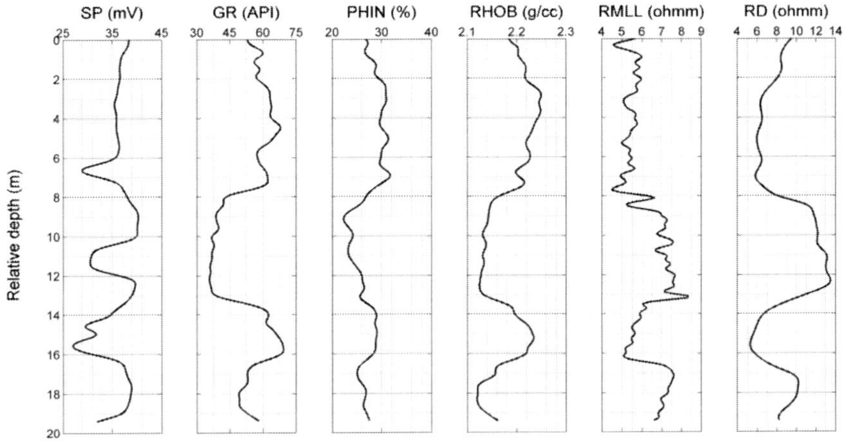

Figure 5. The well logs measured in Well-1 as input for the GA-based interval inversion procedure. Denotations: spontaneous potential (SP), natural γ-ray intensity (GR), density (RHOB), neutron porosity (PHIN), microlaterolog (RMLL), deep resistivity (RD).

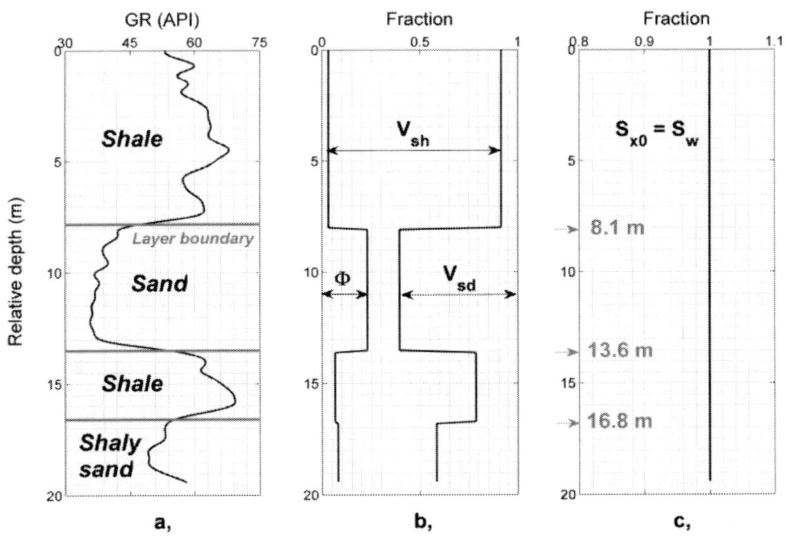

Figure 6. The results of the GA-based interval inversion procedure in Well-1: a, natural γ-ray log (GR) with lithology and manually traced out layer-boundaries; b, estimated porosity (Φ), shale volume (V_{sh}), sand volume (V_{sd}); c, water saturation in the flushed (S_{x0}) and undisturbed zone (S_w), automatically estimated layer-boundary coordinates (indicated by arrows).

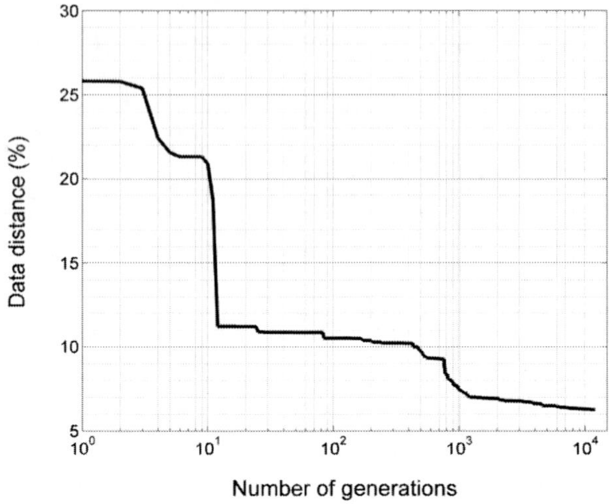

Figure 7. The convergence plot of the GA-based interval inversion procedure in Well-1: data distance of the best individual vs. iteration step.

At the end of the parameter search, a satisfactory fit in data space was achieved as the relative data distance was D_d=6.3%. (The model distance is impossible to compute for a lack of an exact model.)

The inversion results are in Figure 6/b-c. Among the shaly sandy layers there is a productive water-bearing reservoir with high porosity and permeability (second layer). The automatically estimated layer thicknesses are 8.1m, 5.5m, 3.2m and 2.6m, in close agreement with manual results. Since the pores are fully saturated with water along the entire interval, then $S_{x0}=S_w=1$. Shale particles dispersed in the pore space have large specific surface areas, which can absorb a high amount of water. Another type of clay incorporated in the rock matrix may also contain water in their mineral structure. Therefore, along the intervals of shale layers, a 100% water saturation is indicated.

5.2. Application to Hydrocarbon Exploration

The feasibility of the GA-based interval inversion method was tested by using another real well-logging data set originated from a deep Hungarian hydrocarbon exploratory borehole (Well-2). The aim of the interpretation was to find and locate possible hydrocarbon reservoirs and estimate their storage capacity. We processed a 35-meter long interval of a hydrocarbon zone, where the sampling interval of well logging was 0.1 m. The inverted open-hole logs are in Figure 8.

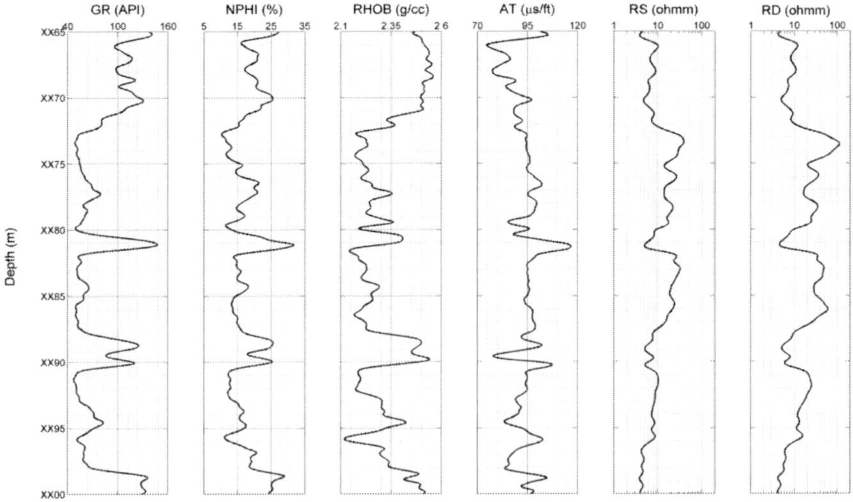

Figure 8. The well logs measured in Well-2 as input for the GA-based interval inversion procedure. Denotations: natural γ-ray intensity (GR), neutron porosity (NPHI), density (RHOB), acoustic traveltime (AT), shallow resistivity (RS), deep resistivity (RD).

Seven homogeneous formations were assumed by a preliminary study of well logs. The sequence consisted of gas-bearing shaly-sand formations with interbedded shale layers. The GA search required 10,000 generations for updating 30 individuals. The control parameters of genetic operators (3.3), (3.6), (3.7) were the probability of selecting the best individual (p_b=0.08), crossover retry (150) and mutation probability (p_m=0.03). The convergence of the inversion procedure was smooth and steady, as it is seen in Figure 9. The optimal solution was found at relative data distance D_d=7.2%.

The logs of estimated petrophysical parameters are in Figure 10/b-c. The method distinguished permeable and impermeable intervals and gave an estimate for rock interfaces at nearly the same places as they were inferred from the *GR* log.

Along the processed interval the evaluation revealed three hydrocarbon reservoirs with high porosity separated by shale beds (Figure 10/a). The estimated thicknesses of the layers were 7.3m, 7.8m, 1.6m, 6.0m, 2.0m and 8.1m.

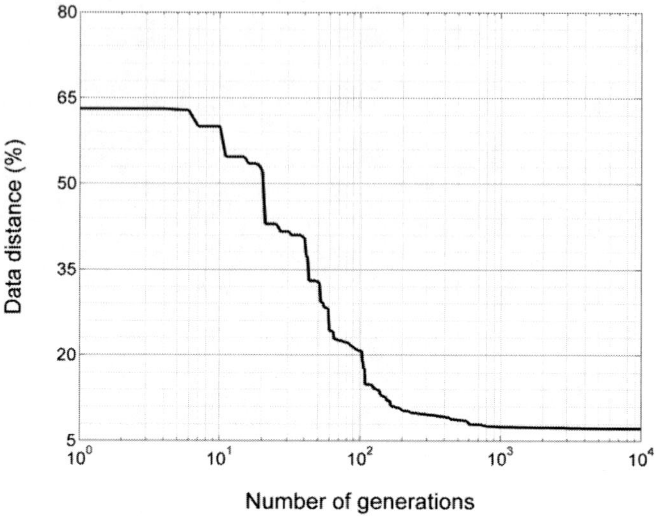

Figure 9. The convergence plot of the GA-based interval inversion procedure in Well-2: data distance of the best individual vs. iteration step.

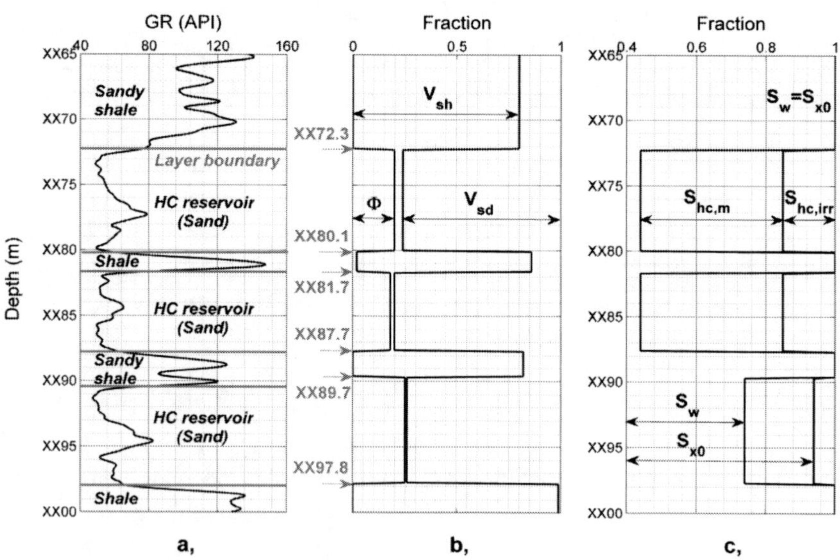

Figure 10. The GA-based interval inversion results in Well-2: a, natural γ-ray log with lithology and manually located layer-boundaries; b, estimated porosity (Φ), shale volume (V_{sh}), sand volume (V_{sd}) and automatically estimated layer-boundary coordinates (indicated by arrows); c, water saturation in the flushed (S_{x0}) and undisturbed zones (S_w), movable hydrocarbon saturation ($S_{hc,m}$), irreducible hydrocarbon saturation ($S_{hc,irr}$).

The movable and irreducible gas saturation based on equations (4.1) and (4.2) were approximately 41% and 15% (first and second hydrocarbon reservoirs), 21% and 7% (third hydrocarbon reservoir), respectively.

The estimation of permeability as a derived parameter from inverse modeling was also made by using an empirical formula including porosity and other properly chosen constants (Timur, 1968). We calculated permeability values from 1,300 to 1,900 mDarcies in the porous and permeable rocks. It was concluded that the results of interpretation showed good quality hydrocarbon reservoirs.

CONCLUSION

In this chapter, the Float-Encoded Genetic Algorithm was presented as an effective global optimization tool used for the inversion processing of well-logging data. The interval inversion method gives an estimate for petrophysical parameters and the coordinates of formation boundaries at arbitrary depth intervals in a joint inversion procedure. The highly over-determined inverse problem results in an accurate, reliable and robust solution for the logged interval instead of making a prediction in separate measuring points. As the procedure of forward modeling is not very time-consuming, the CPU times take only several minutes by using a quad-core processor-based workstation, which can be further reduced by applying a subsequent combination of GA and a fast linear optimization method.

The purport of the interval inversion technique is to use series expansion for discretizing the model parameters. Assuming homogeneous layers, the interval inversion method enables a more accurate and reliable separation of movable and irreducible hydrocarbon volumes than by conventional (local) inversion methods. In the case of inhomogeneous layers, a higher resolution of petrophysical parameters can be achieved by using power or other appropriate types of basis functions. The interval inversion method can make an automatic prediction of layer boundary coordinates. The effective layer thickness is an important parameter in calculating the producible water or hydrocarbon reserves. The capability of giving an automatic estimate to geometrical quantities sets new perspectives to describe the morphology of geological structures in more detail. The interval inversion algorithm can be developed for multi-well applications by using basis functions depending on more spatial coordinates. In that case, the formation boundaries can be approximated by appropriately chosen two or three dimensional basis functions. Thus, the surface and the volume of rock structures can be determined automatically. The method is supported by large amounts of wellbore information and large extents of over-determination. The latter is not reduced significantly by introducing some additional unknowns into the inverse problem. Therefore, some other zone parameters, routinely treated as local specific constants, can be estimated within the inversion procedure. As a result, the areal distributions of petrophysical parameters can be given from a more objective source that supports all fields of Earth science with useful *in-situ* information.

ACKNOWLEDGMENTS

The described work was carried out as part of the TÁMOP-4.2.2/B-10/1-2010-0008 project in the framework of the New Hungary Development Plan. The realization of this project is supported by the European Union, co-financed by the European Social Fund. The first author thanks to the support of the János Bolyai Research Fellowship of the Hungarian Academy of Sciences. The second author is grateful for the support of the MTA-ME Research Group for Earth Sciences.

REFERENCES

Alberty, M.; Hashmy, K. *SPWLA Symposium Transactions*. 1984, Z, 1-17.

Asquith, G.; Krygowski, D. Basic well log analysis; AAPG Methods in Exploration 16; *American Association of Petroleum Geologists*, Tulsa, 2004.

Ball, S. M.; Chace, D. M.; Fertl, W. H. Proc. *SPE Eastern Regional Meeting*. 1987, 17034, 61-85.

Dobróka, M. Scientific Report for the Hungarian Oil and Gas Company. 1993, *University of Miskolc.*

Dobróka, M.; Szabó, N. P. Geosciences, *Publications of the University of Miskolc*. 2001, 59, 115-137.

Dobróka, M.; Szabó, N. P. *Acta Geodaetica et Geophysica Hungarica*. 2005, 40, 203-214.

Dobróka, M.; Szabó, P. N.; Kiss, B.; Krasznavölgyi, *Proc. 67th EAGE Conference and Exhibition.* 2005, 215, 1-4.

Dobroka, M.; Kiss, B.; Szabo, N.; Toth, J.; Ormos, T. *Proc. 69th EAGE Conference and Exhibition*. 2007, 092, 1-5.

Dobróka, M.; Szabó, P. N.; Cardarelli, E.; Vass, P. *Acta Geodaetica et Geophysica Hungarica*. 2009, 44, 459-479.

Dobróka, M.; Szabó, N. P. Acta Geophysica. 2011, 59, 907-934.

Dobróka, M.; Szabó, N. P.; Turai, E. *Acta Geodaetica et Geophysica Hungarica*. 2012, 47, 172-184.

Dobróka, M.; Szabó, N. P. *Journal of Petroleum Science and Engineering*. 2012, 86–87, 144-152.

Dolton, G. L. USGS Bulletin 2204–B. 2006, 1-47.

Holland, J. H. Adaptation in natural and artificial systems; Univ. Michigan Press, *Ann Arbor*, 1975.

Houck, C. R.; Joines, J. A.; Kay M. G. *Technical report.* 1995, 1-14.

Mayer, C.; Sibbit, *A. Proc. SPE Annual Fall Technical Conference and Exhibition.* 1980, 9341, 1-14.

Menke, W. Geophysical data analysis: discrete inverse theory; Academic Press, New York, 1984.

Metropolis, N.; Rosenbluth, A.; Rosenbluth, M.; Teller, A.; Teller, E. *J. Chem. Phys.* 1953, 21, 1087–1092.

Michalewicz, Z. Genetic Algorithms + Data Structures = Evolution Programs; Springer-Verlag, New York, 1992.

Poupon, A.; Leveaux, J. *The Log Analyst*. 1971, 12, 3-8.

Sen M.; Stoffa P. L. In Seismic Exploration CD-ROM; Helbig, K.; Treitel, S.; Elsevier Science, 1997.

Serra, O. Fundamentals of well-log interpretation; Elsevier, Amsterdam, 1984.

Szabó, P. N.; Dobróka, M.; Hursán, L. Proc. *65th EAGE Conference and Exhibition*. 2003, 261, 1-4.

Szabó, N. P. *Geophysical Transactions*. 2004, 44, 313-329.

Tarantola, A. Inverse problem theory and methods for model parameter estimation; *Society for Industrial and Applied Mathematics*, Philadelphia, 2005.

Telford, W. M.; Geldart, L. P.; Sheriff, R. E. Applied Geophysics; Cambridge University Press, 1990.

Timur, A. *The Log Analyst.* 1968, 9, 8-17.

In: Global Optimization
Editor: Angelika Michalski

ISBN: 978-1-62417-796-5

Chapter 4

THE PARTICLE COLLISION ALGORITHM: A METROPOLIS OPTIMIZATION METHOD

Wagner F. Sacco[1*]***,***
Anderson Alvarenga de Moura Meneses[1]***,***
Ana Carolina Rios-Coelho[1]
and Nélio Henderson[2]
[1]Universidade Federal do Oeste do Pará, Brazil
[2]Instituto Politécnico da Universidade do Estado do Rio de Janeiro, Brazil

ABSTRACT

Stochastic optimization methods based on the simulated annealing paradigm have been actively developed in the last thirty years and successfully applied to many complex optimization problems. These methods though very powerful are not free from practical drawbacks, the main one being that performance is too sensitive to the choice of free parameters. Ideally, an optimization algorithm should not rely on user-defined or problem-dependent parameters, except a stopping criterion. One such algorithm, that was recently introduced and applied to real-world problems, is reviewed here. Denoted the "Particle Collision Algorithm" (PCA), the algorithm is loosely inspired by the physics of nuclear particle collision reactions, particularly scattering and absorption. These events promote the exploration of the search space and the exploitation of its most promising areas. The PCA resembles in its structure that of simulated annealing and can also be considered a

Metropolis algorithm, as a worse trial solution can be accepted with a certain probability. This acceptance may avoid the convergence to local optima. Since its introduction, some variants of the original algorithm have been created, and it has also been hybridized with other techniques. The PCA and its derivations have been successfully applied to real-world optimization problems, such as a nuclear reactor design optimization problem, radiative transference and pollutant localization inverse problems, chemical equilibrium nonlinear systems, and course-timetabling problems.

Keywords: Global optimization, metropolis algorithm, stochastic optimization

1. INTRODUCTION

Stochastic optimization methods based on the simulated annealing paradigm [1] have been actively developed in the last thirty years and successfully applied to many complex optimization problems. These methods though very powerful are not free from practical drawbacks, the main one being that performance is too sensitive to the choice of free parameters, such as, for example, the annealing schedule and initial temperature [2].

Ideally, an optimization algorithm should not rely on user-defined or problem-dependent parameters and should not converge to a suboptimal solution. Given the appropriate annealing schedule, simulated annealing is guaranteed to converge to the global optimum [3], as it is a Metropolis-based algorithm [4], where a worse solution can be accepted with a certain probability. However, its rate of convergence is strongly dependent on the user-specified initial parameters. So, the best outcome would be the development of an algorithm similar to simulated annealing, but without the burden of parameter specification.

One such algorithm, which does not rely on user-supplied parameters to perform the optimality search, except a stopping criterion, is reviewed here. Its structure was outlined, along with some preliminary results, by Sacco and de Oliveira [5], and it was introduced in detail by Sacco et al. [6]. Named the "Particle Collision Algorithm" (PCA), the algorithm is loosely inspired by the physics of nuclear particle collision reactions [7], particularly scattering and absorption. Thus, a particle that hits a high-fitness "nucleus" would be "absorbed" and would explore the boundaries. On the other hand, a particle that hits a low-fitness region would be scattered to another region. This permits us to simulate the exploration of the search space and the exploitation of the most promising areas of the fitness landscape through successive scattering and absorption collision events.

In this chapter, we present a review of the PCA, including extensions of the original algorithm and applications. The remainder of this chapter is organized as follows. We present the canonical version of the algorithm in detail in section 2. In the following section, some applications of the algorithm to real-world problems are described. In section 4, some variants of the original algorithm that have been proposed in the literature are presented. In section 5, hybrid algorithms involving the PCA are outlined. Finally, in section 6, the concluding remarks are made.

2. The Canonical PCA

The PCA algorithm resembles in its structure that of simulated annealing: first an initial configuration is chosen; then there is a modification of the old configuration into a new one. The qualities (i.e., objective-function values) of the two configurations are compared. A decision is then made on whether the new configuration is "acceptable". If it is, it serves as the old configuration for the next step. If it is not acceptable, the algorithm proceeds with a new change of the old configuration. The PCA can also be considered a Metropolis algorithm, as a trial solution can be accepted with a certain probability. This acceptance may avoid the convergence to local optima.

A summary of the PCA is given below [8]. The algorithm's default is for maximization problems. For minimization problems, just multiply the objective-function by -1 and invert the ratio in $p_{scattering}$.

Initialization Step

Generate an initial random solution **OldConfig**.

Main Step

Do until termination criterion is reached:

1. Generate a stochastic perturbation **NewConfig** of the solution **OldConfig**:
   NewConfig ← **OldConfig**+{[(**Upper**-**OldConfig**)* rand (0,1)]-[(**OldConfig**-**Lower**)*(1-rand (0,1))]}
2. If Quality(**NewConfig**) >Quality(**OldConfig**), let
   OldConfig ← **NewConfig** and go to *Exploitation*. Otherwise, go to *Scattering*.

End-Do

Exploitation

Do for n iterations:

1. Generate a stochastic perturbation **NewConfig** of the solution **OldConfig**:
   NewConfig ← **OldConfig**+{[(**Upper**-**OldConfig**)* rand (0,1)]-[(**OldConfig**-**Lower**)*(1-rand (0,1))]};
   Upper ← rand (1.0,1.2)***OldConfig**;
   Lower ← rand (0.8,1.0)***OldConfig**;
   Check if **Upper** and **Lower** are within the boundaries;
   If not, make equal to upper and lower boundary, respectively.
2. If Quality(**NewConfig**) >Quality(**OldConfig**), let
   OldConfig ← **NewConfig** .

End-Do
return

Scattering

1. Calculate $p_{scattering} = 1 - $ (Quality(**NewConfig**) / **CurrentBest**).
2. If $p_{scattering} >$ rand(0,1), let **OldConfig** receive a random solution. Otherwise, go to *Exploitation*.

return

The termination criterion is usually a maximum number of objective-function evaluations.

The "stochastic perturbation" in the beginning of the loop consists in random variations in each variable's values within their ranges. Note that in PCA the perturbation mechanism is different from simulated annealing, where generally only a few variables are perturbed at a time [9].

If the quality or fitness of the new configuration is better than the fitness of the old one, then the "particle" is "absorbed", there is an exploitation of the boundaries searching for an even better solution. Function "Exploitation" performs this local search, generating a small stochastic perturbation of the solution inside a loop, where the number of iterations is usually set as one-hundred [8]. The "small stochastic perturbation" is similar to the previous stochastic perturbation, but each variable's new value is kept within the boundaries of the original value.

Otherwise, if the quality of the new configuration is worse than the old configuration's, the "particle" is "scattered". By scattering we mean that the new configuration receives random values between the lower and upper bounds of each variable. The scattering probability $p_{scattering}$ is inversely proportional to its fitness function value. Thus, a low-fitness particle will have a greater scattering probability. This mechanism is essential to the success of the algorithm, as it introduces a purely random component, which is responsible to avoid getting stuck at local optima.

Regarding the *Scattering* routine, we have the attribution

$$p_{scattering} = 1 - \frac{\text{Quality}(\mathbf{NewConfig})}{\mathbf{CurrentBest}} \tag{1}$$

It is quite clear that **CurrentBest** may assume very small values, leading to a division by zero and to an infinite value for the fractional term. To remedy this situation, we created a routine called *Divide-Check*, which checks the occurrence of a division by zero. In case it occurs, $p_{scattering}$ receives the following value:

$$p_{scattering} = -\text{signal}[\text{Quality}(\mathbf{NewConfig})](\mathit{MaxMach}), \tag{2}$$

where *MaxMach* is the magnitude of the largest number representable by the machine under use. Note that

$$\text{signal}[x] = \begin{cases} +1, \text{if } x \geq 0, \\ -1, \text{if } x < 0. \end{cases} \tag{3}$$

3. Applications of the PCA in the Literaure

3.1. Nuclear Engineering Problems

As the creators of the PCA, Sacco and de Oliveira, are nuclear engineers, the first applications of the algorithm were in their field of expertise.

Sacco et al. [6] applied the PCA to a nuclear reactor core design optimization problem [10]. The detailed description of this problem is beyond the scope of this chapter (for more details, see [11]). Briefly stated, it consists in adjusting several reactor cell parameters, such as dimensions, enrichment and materials, in order to minimize the average peak-factor in a three-enrichment-zone reactor, considering restrictions on the average thermal flux, criticality and sub-moderation. The PCA outperformed the established simulated annealing and genetic algorithm [12] methods, obtaining better objective-function values with less computational effort.

Another application of the algorithm in the field of nuclear engineering was to a real-world problem that consists in finding the optimal surveillance tests policy of an operating nuclear power plant auxiliary feedwater system, which is expressed in terms of average system unavailability [13]. Employing the genetic algorithm and variations [13-15], this problem took hours to be solved, with an order of magnitude of 10^6 objective-function evaluations. The PCA obtained the best average system unavailability so far, in 10^5 objective-function evaluations [16], or one-tenth of the computational effort required by the previous efforts.

3.2. Inverse Problems

Knupp et al. [17] successfully employed the PCA to an inverse radiative heat transfer problem, where the aim is the estimation of the optical thickness, single scattering albedo and diffuse reflectiveness of one-dimensional homogeneous participating media [18].

Due to its simplicity, Fernandes et al. [19] chose the PCA to solve an inverse problem of damage assessment in an Euler-Bernoulli beam within the framework of the wave propagation approach. The experiments were performed using slender aluminum beams with imposed damage scenarios. The cross section area assessment was provided by the Particle Collision Algorithm. The authors report that this method presented a noticeable sensitivity with respect to the damage identification [19].

3.3. Course-Timetabling Problems

The university course-timetabling problem involves assigning a set of courses, lectures and students to a specific number of rooms and timeslots. The goal is to produce a high-quality timetable that satisfies all hard constraints and attempts to accommodate soft constraints as much as possible [20]. Abuhamdah and Ayob [20] used the PCA to solve a benchmark consisting of eleven datasets introduced by Socha et al. [21]. The problems involve a set of events to be scheduled in forty-five timeslots, a set of rooms where the events take place, a set of students who attend these events, and a set of features characterizing the rooms and required by the events [20]. The PCA was compared against simulated annealing and another Metropolis algorithm called the Great Deluge Algorithm [22]. The PCA demonstrated to be competitive, significantly outperforming its competitors in some instances.

4. Variants of the PCA

In this section, some variants of the original PCA are presented, following three axes: better local search mechanisms, multiple-particle strategies, and a problem-specific variation.

4.1. Other Local Searching Schemes

Sacco et al. [6] pointed out in the conclusion of their article on the canonical PCA that the local search mechanism (exploitation phase in the original algorithm) could be improved. Recently, the exploitation phase has been replaced by three well-known deterministic local search algorithms:

Levenberg-Marquadt [23], Nelder-Mead Simplex [24], and Hooke-Jeeves [25]. The schemes always follow the pattern given below:

Initialization Step

Generate an initial random solution **OldConfig**.

Main Step

Do until termination criterion is reached:

1. Generate a stochastic perturbation **NewConfig** of the solution **OldConfig**:

   NewConfig ← **OldConfig**+{[(**Upper-OldConfig**)* rand (0,1)]-[(**OldConfig-Lower**)*(1-rand (0,1))]}
2. If Quality(**NewConfig**) >Quality(**OldConfig**), let **OldConfig** ← **NewConfig** and go to *DLSA*. Otherwise, go to *Scattering*.

End-Do

Deterministic Local Search Algorithm (DLSA)

1. Apply the local search algorithm using as staring point the solution **OldConfig**, generating a solution **NewConfig**;
2. If Quality(**NewConfig**) >Quality(**OldConfig**), let **OldConfig** ← **NewConfig** .

return

Scattering

1. Calculate $p_{scattering} = 1 - $(Quality(**NewConfig**) / **CurrentBest**).
2. If $p_{scattering} >$ rand(0,1), let **OldConfig** receive a random solution. Otherwise, go to *DLSA*.

return

In order to solve the problem previously attacked in ref. [17], Knupp et al. [26] employed a PCA with local searching phase performed by the Levenberg-Marquadt (LM) algorithm, obtaining better results than the original PCA. The LM algorithm is a gradient-based method derived from the method of Gauss-Newton. For a concise description, see, for example, [27].

One must bear in mind, however, that the LM algorithm requires the computation of derivatives, which is not only computationally demanding, but restricts its application, as there is a wide range of problems that cannot be handled using gradient-based methods. This motivated the use of direct search methods [28], which are derivative-free.

Sacco et al. [29] applied the PCA with local search performed by the most popular direct search method, Nelder-Mead Simplex, to the nuclear reactor core design optimization problem. This problem cannot be solved using gradient-based techniques, hence the necessity of using a direct search technique. The results obtained were better, more consistent, and requiring less computational effort than those achieved using the canonical PCA. Nelder-Mead Simplex uses an $n+1$ dimensional simplex to solve an $n-$dimensional problem through the operations of reflection, expansion, contraction, and shrinkage of this polytope. For further details, see [30].

However, Nelder-Mead Simplex is not robust. It may work better than any direct search method in one problem and fail in the next one [28]. Rios-Coelho el al. [8] replaced Nelder-Mead Simplex with the method of Hooke-Jeeves (HJ), which is robust and easy to implement.

Also known as pattern search, the Hooke-Jeeves algorithm starts from an initial base point and explores each coordinate axis with its own step size. In an $n-$dimensional problem, trial points in all n positive and negative coordinate directions are compared and the best point is found. If the best new trial point is better than the base point, then an attempt is made to make another move in the same direction, since the step from these points was a good one. If, however, none of the trial points improve the original point, the step is presumed to have been too large, so the procedure repeats with smaller step sizes. For a fine exposition of the HJ algorithm, see [31].

Rios-Coelho et al. [8] applied the PCA with local search by HJ (HJPCA) to a set of global optimization test functions widely employed in the literature (see, for example, [32-35]) and to a real-world chemical equilibrium problem [36] that has been a great challenge for nonlinear-system solvers [37-39].

The set of test functions contains functions with multiple global optima, with an isolated global optimum, and with high dimensionality, being, thus, a rigorous testbed for global optimization algorithms. When applied to this testbed, HJPCA outperformed two state-of-art global optimization methods, GLOBAL [33] and C-GRASP [35].

The chemical equilibrium problem concerns the combustion of propane in the air to form ten products. This chemical reaction generates a system of ten equations in ten unknowns, which can be reduced to a system of five equations in five unknowns. Rios-Coelho et al. [8] solved both systems formulating them as optimization problems. HJPCA found the root of the 5×5 system in

$100/100$ runs, taking an average of 0.522s. C-GRASP took an average of 37.53s with a $10/10$ rate of success [34]. In the 10×10 system, HJPCA was successful in $96/100$ runs taking an average of 3.093s, while C-GRASP obtained a rate of $8/10$ taking in the average 201.58s [34]. Of course, time is not the best performance measure, as it is machine dependent. However, the results in the literature for this kind of problem are all in terms of time (see, for example, [36-39]).

4.2. Multiple-Particle Variants

In order to promote a greater exploration of the search space and eventually parallelize the algorithm, Sacco et al. [40] introduced a Populational PCA (PopPCA), which is a multiple-particle version of the PCA that incorporates some features of the genetic algorithm. Exactly like the standard PCA, each particle moves in the search space through successive collision events. By the end of a generation, though, the particles interact for reproduction. The fittest individuals survive and constitute the population for the next generation.

A summary of PopPCA is given below. In relation to the pseudo code presented in section 2, the difference is the introduction of a loop to iterate over the population and of function "*New Population*", which generates the new population for the next generation in a fashion similar to the genetic algorithm [41].

Initialization Step

For $i = 1$ to $PopSize$

Generate an initial random solution $\mathbf{OldConfig}[i]$.

End-For

Main Step

Do until termination criterion is reached:

For $i = 1$ to $PopSize$

1. Generate a stochastic perturbation **NewConfig**[*i*] of the solution **OldConfig**[*i*]:
   NewConfig[*i*] ← **OldConfig**[*i*]+{[(**Upper**-**OldConfig**[*i*])* rand (0,1)]-[(**OldConfig**[*i*]-**Lower**)*(1-rand (0,1))]}

2. If Quality(**NewConfig**[*i*]) >Quality(**OldConfig**[*i*]), let **OldConfig**[*i*] ← **NewConfig**[*i*] and go to *Exploitation*. Otherwise, go to *Scattering*.

End-For
Go to *New Population*.
End-Do

Exploitation

Do for *n* iterations:

1. Generate a stochastic perturbation **NewConfig**[*i*] of the solution **OldConfig**[*i*]:
   NewConfig[*i*] ← **OldConfig**[*i*]+{[(**Upper**-**OldConfig**[*i*])*rand (0,1)]-[(**OldConfig**[*i*]-**Lower**)*(1-rand (0,1))]};
   Upper ← rand (1.0,1.2)***OldConfig**[*i*];
   Lower ← rand (0.8,1.0)***OldConfig**[*i*];
   Check if **Upper** and **Lower** are within the boundaries;
   If not, make equal to upper and lower boundary, respectively.
2. If Quality(**NewConfig**[*i*]) >Quality(**OldConfig**[*i*]), let **OldConfig**[*i*] ← **NewConfig**[*i*].

End-Do
return

Scattering

1. Calculate $p_{scattering} = 1 -$ (Quality(**NewConfig**[*i*]) / **CurrentBest**).
2. If $p_{scattering} >$ rand(0,1), let **OldConfig**[*i*] receive a random solution. Otherwise, go to *Exploitation*.

return

New Population

1. Cross two random individuals from the population;
2. Rank parents and siblings in descending fitness order;

3. Pick up the first *PopSize* individuals from the ranked array, filling the array **OldConfig** from $i = 1$ to *PopSize.*

return

At the end of a generation, function "*New Population*" is called, randomly pairing and crossing the individuals using "one-point crossover" [41], where a cutting point is determined at random and the segments before this point are swapped. After that, the parents and siblings are ranked in descending fitness order and the first *PopSize* individuals from this ordered array survive to the next generation.

The PopPCA was applied to the nuclear reactor core design problem described in section 3.1, obtaining, in the average of ten experiments, a better result than PCA's.

Recently, da Luz and collaborators [42] introduced a simple, yet ingenuous multiple-particle variety of the PCA: the Multiple-Particle Collision Algorithm (denoted as MPCA). The authors had a twofold motivation: to promote a better exploration of the search space, avoiding premature convergence to a local optimum, and to create a method easy to implement in a parallel environment, leading to a great reduction of computing time [42].

In the MPCA, the particles interact with each other sharing information about the best configuration found so far. This is performed through a blackboard strategy. The pseudo code of the algorithm is given below.

Initialization Step

For $i = 1$ to *PopSize*

Generate an initial random solution **OldConfig**[i].

End-For

Update *Blackboard*

Main Step

For $i = 1$ to *PopSize*

Do until termination criterion is reached:

1. Update *Blackboard*

2. Generate a stochastic perturbation **NewConfig**[i] of the solution **OldConfig**[i]:
   NewConfig[i] ← **OldConfig**[i]+{[(**Upper**-**OldConfig**[i])*rand(0,1)]-[(**OldConfig**[i]-**Lower**)*(1-rand(0,1))]}

3. If Quality(**NewConfig**[i]) >Quality(**OldConfig**[i]), let **OldConfig**[i] ← **NewConfig**[i] and go to *Exploitation*. Otherwise, go to *Scattering*.

End-Do
End-For

Exploitation

Do for n iterations:

1. Generate a stochastic perturbation **NewConfig**[i] of the solution **OldConfig**[i]:
   NewConfig[i] ← **OldConfig**[i]+{[(**Upper**-**OldConfig**[i])*rand(0,1)]-[(**OldConfig**[i]-**Lower**)*(1-rand(0,1))]};
   Upper ← rand(1.0,1.2)***OldConfig**[i];
   Lower ← rand(0.8,1.0)***OldConfig**[i];
   Check if **Upper** and **Lower** are within the boundaries;
   If not, make equal to upper and lower boundary, respectively.
2. If Quality(**NewConfig**[i]) >Quality(**OldConfig**[i]), let **OldConfig**[i] ← **NewConfig**[i].

End-Do
return

Scattering

1. Calculate $p_{scattering} = 1 - (\text{Quality}(\textbf{NewConfig}[i]) / \textbf{CurrentBest})$.
2. If $p_{scattering} > \text{rand}(0,1)$, let **OldConfig**[$i$] receive a random solution. Otherwise, go to *Exploitation*.

return

Note that in the inner loop of the *Main Step*, the termination criterion must be $1/PopSize$ of the conventional PCA's, so that both algorithms have the same total computational cost. One must remember, however, that in the case of MPCA, this total cost is split among multiple processors. The MPCA promotes a time reduction from PCA's t to $t / n_{processors}$.

The MPCA was implemented by da Luz et al. [42] using MPI libraries in a multiprocessor architecture with distributed memory. They solved two inverse problems with this algorithm: a radiative transference problem similar to that presented in ref. [18] and the localization of polluting sources, where the particle dispersion was modeled using a Lagrangian Model for Buoyant Dispersion in Atmosphere (LAMBDA) [42].

4.3. Multi-Neighborhood PCA

After their work with the PCA (section 3.3), Abuhamdah and Ayob conceived a problem-specific version of the PCA, called Multi-Neighborhood PCA, in order to solve course-timetabling problems [43]. The authors incorporated in the PCA some heuristics to produce feasible timetables [44], which include neighborhood structure moves, therefore the name of the algorithm. These heuristics were employed in step (1) of the *Main Step* and in the *Exploitation* routine of the PCA.

The results obtained were better than those achieved by the conventional PCA [43]. Moreover, this PCA variant obtained, in that occasion, the best result in the literature for one of the datasets [43].

5. HYBRID APPROACHES

Besides being modified, in recent years, the PCA has also been hybridized with other algorithms.

The first hybrid scheme was proposed by Knupp et al. [17]. These authors combined the PCA with the Luus-Jaakola algorithm [45], which is a random search algorithm that has been applied mostly in the field of Chemical Engineering [46-48]. The PCA was connected to the LJ algorithm in order to generate an initial guess for the latter. This hybrid approach was 26% faster than the conventional LJ, taking an average of $1\,\text{h}\,55\,\text{min}$ to solve the inverse problem briefly described in section 3.2 [17].

To the best of our knowledge, the other hybrid approach using the PCA that has been published so far is that introduced by Lei et al. [49]. This algorithm was used to solve an inverse problem: image reconstruction for electrical capacitance tomography. As the traditional homotopy technique [50] was converging to a local optimum, the authors decided to combine a homotopy algorithm, the populational algorithm Multi-Harmony Search algorithm (MHS) [51], and the PCA in the following way [49]:

Step 1. An initial solution is given and the homotopy algorithm is implemented to obtain a local optimal solution;
Step 2. Generate the initial solution swarms in the neighborhood of the local optimal solution obtained in Step 1, and run the MHS algorithm to obtain a new solution;
Step 3. Set the best solution that is obtained by the MHS algorithm into the initial value of the PCA and run this algorithm to obtain a new solution;
Step 4. Set the solution that the PCA finds as an initial value, looping to Step 1; repeat the above process until a satisfactory solution is obtained.

This algorithm was tested in six typical reconstruction objects using a twelve-electrode square sensor [49]. According to the authors, the results show the efficiency and superiority of this hybrid approach in solving the electrical capacitance tomography image reconstruction problem [49].

CONCLUSION

The Particle Collision Algorithm was created as an alternative to Simulated Annealing, whose effectiveness strongly depends on parameter specification [2]. Since its introduction [5], the PCA has demonstrated to be a robust and efficient method, besides being easy to implement. The results obtained in practical problems [6,8,16,20] corroborate not only these points, but also demonstrate PCA's competitiveness against some established algorithms like the genetic algorithm and simulated annealing, and even the state-of-art techniques GLOBAL and C-GRASP.

The variants with local search performed by deterministic algorithms performed even better than the canonical version, especially the one that uses the method of Hooke and Jeeves [8].

Another improvement of the original algorithm is the multiple-particle version proposed by da Luz et al. [42], the MPCA, which besides reducing computing time in a parallel implementation, has the potential to promote a greater exploration of the search space.

A MPCA with local search performed by the Hooke-Jeeves algorithm would incorporate the best features of both PCA variants, with the prospect of being a robust multi-purpose algorithm.

Acknowledgments

W.F.S. and N.H. gratefully acknowledge the financial support provided by CNPq (Conselho Nacional de Desenvolvimento Científico e Tecnológico, Ministry of Science & Technology, Brazil). Portions of this chapter have been published in the journals Applied Mathematics and Computation, Annals of Nuclear Energy, Progress in Nuclear Energy, and The Imaging Science Journal.

References

[1] Kirkpatrick, S.; Gelatt, C.D.;Vecchi, M.P. Optimization by Simulated Annealing. *Science*, 1983, 220, 671-680.

[2] Carter, J.N. Genetic algorithms for incore fuel management and other recent developments in optimization. *Advances in Nuclear Science and Technology,* 25, 113-154, Plenum Press, New York, 1997.

[3] Aarts, E.; Korst, J. *Simulated Annealing and Boltzmann Machines*. John Wiley and Sons, New York, 1989.

[4] Metropolis, N., Rosenbluth, A.W., Rosenbluth, M.N., Teller, A.H., Teller, E., 1953. Equations of state calculations by fast computing machines. *Journal of Chemical Physics* 21, 1087-1092.

[5] Sacco, W.F.; de Oliveira, C.R.E. A New Stochastic Optimization Algorithm based on Particle Collisions. *Transactions of the American Nuclear Society*, 2005, 92, 657-659.

[6] Sacco, W.F.; de Oliveira, C.R.E.; Pereira, C.M.N.A. Two stochastic optimization algorithms applied to nuclear reactor core design. *Progress in Nuclear Energy,* 2006, 48, 525-539.

[7] Duderstadt, J.J.; Hamilton, L.J. *Nuclear Reactor Analysis*. John Wiley and Sons, New York, 1976.

[8] Rios-Coelho, A.C.; Sacco, W.F.; Henderson, N. A Metropolis algorithm combined with Hooke-Jeeves local search method applied to global optimization. *Applied Mathematics and Computation,* 2010, 217, 843-853.

[9] Brooks, S.P.; Morgan, B.J.T. Optimization using simulated annealing. *Statistician*, 1995, 44, 241–257.

[10] Pereira, C.M.N.A.; Schirru, R.; Martinez, A.S. Basic investigations related to genetic algorithms in core designs. *Annals of Nuclear Energy,* 1999, 26, 173-193.

[11] Sacco, W.F.; Meneses, A.A.M.; Pereira, C.M.N.A. Global Optimization Methods Applied to the Nuclear Reactor Core Design. *Advances in Engineering Research,* 3, 505-519, Nova Publishers, New York, 2011.

[12] Holland, J.H. *Adaptation in Natural and Artificial Systems.* University of Michigan Press, Ann Arbor, 1975.

[13] Lapa, C.M.F.; Pereira, C.M.N.A.; Frutuoso e Melo, P.F. An application of genetic algorithms to surveillance tests optimization of a PWR auxiliary feed-water system. *International Journal of Intelligent Systems,* 2002, 17, 813-831.

[14] Pereira, C.M.N.A.; Lapa, C.M.F. Parallel island genetic algorithm applied to a nuclear power plant auxiliary feedwater system surveillance tests policy. *Annals of Nuclear Energy,* 2003, 30, 1665-1675.

[15] Sacco, W.F.; Lapa, C.M.F.; Pereira, C.M.N.A.; de Oliveira, C.R.E. A niching genetic algorithm applied to a nuclear power plant auxiliary feedwater system surveillance tests optimization. *Annals of Nuclear Energy,* 2006, 33, 753-759.

[16] Sacco, W.F.; Lapa, C.M.F.; Pereira, C.M.N.A.; Alves Filho, H. A Metropolis algorithm applied to a nuclear power plant auxiliary feedwater system surveillance tests optimization. *Progress in Nuclear Energy,* 2008, 50, 15-21.

[17] Knupp, D.C.; Silva Neto, A.J.; Sacco, W.F. Radiative Properties Estimation with the Luus-Jaakola and the Particle Collision Algorithm. *Computer Modeling in Engineering & Sciences,* 2009, 54, 121-146.

[18] Silva Neto, A.J.; Roberty, N.C.; Pinheiro, R.P.F.; Alvarez Acevedo, N.I. Inverse Problems Explicit and Implicit Formulations with Applications in Engineering, Biophysics and Biotechnology. *Inverse Problems in Science and Engineering,* 2007, 15, 373-411.

[19] Fernandes, K.M.; Sacco, W.F.; Tenenbaum, R.A.; Stutz, L.T.; Silva Neto, A.J.; Soeiro, F.J.C.P. Solution of the Inverse Damage Assessment

Problem with the Particle Collision Algorithm. *8th World Congress on Structural and Multidisciplinary Optimization,* Lisbon, 2009.

[20] Abuhamdah, A.; Ayob, M. Experimental Result of Particle Collision Algorithm for Solving Course Timetabling Problems. *International Journal of Computer Science and Network Security,* 2009, 9, 134-142.

[21] Socha, K.; Knowles, J; Samples, M. A max-min ant system for the university course-timetabling problem. *Lecture Notes in Computer Science,* 2463, 1-13, Springer, Berlin, 2002.

[22] Dueck, G. New optimization heuristics – the Great Deluge Algorithm and Record-to-Record Travel. *Journal of Computational Physics,* 104, 96-92.

[23] Levenberg, K. A Method for the Solution of Certain Non-Linear Problems in Least Squares. *Quarterly of Applied Mathematics,* 1944, 2, 164-168.

[24] Nelder, J.A.; Mead, R. A simplex method for minimization. *Computer Journal,* 1965, 7, 308-313.

[25] Hooke, R.; Jeeves, T.A. "Direct search" solution of numerical and statistical problems. *Journal of the Association for Computing Machinery,* 1961, 8, 212-229.

[26] Knupp, D.C.; Sacco, W.F.; Silva Neto, A.J. Radiative Properties Estimation with a Combination of the Particle Collision Algorithm and the Levenberg-Marquadt Method. *Inverse Problems Symposium,* East Lansing, USA, 2007.

[27] Jang, J.S.R.; Sun, C.T.; Mizutani, E. *Neuro-Fuzzy and Soft Computing – a Computational Approach to Learning and Machine Intelligence.* Prentice Hall, Upper Saddle River, NJ, 1997.

[28] Lewis, R.M.; Torczon, V.; Trosset, M.W. Direct Search Methods: Then and Now. *Journal of Computational and Applied Mathematics,* 2000, 124, 191-207.

[29] Sacco, W.F.; Alves Filho, H.; Henderson, N.; de Oliveira, C.R.E. A Metropolis algorithm combined with Nelder-Mead simplex applied to nuclear reactor core design. *Annals of Nuclear Energy,* 2008, 35, 861-867.

[30] Kelley, C.T. Detection and remediation of stagnation in the Nelder-Mead algorithm using a sufficient decrease condition. *SIAM Journal on Optimization,* 1999, 10, 43, 55.

[31] Bazaraa, M.S.; Sherali, H.D.; Shetty, C.M. *Nonlinear Programming – Theory and Algorithms.* John Wiley and Sons, New York, 1993.

[32] Hedar, A.R.; Fukushima, M. Tabu search directed by direct search methods for nonlinear global optimization. *European Journal of Operational Research,* 2006, 170, 329-349.

[33] Csendes, T.; Pal, L.; Sendín, J.O.H.; Banga, J.R. The GLOBAL optimization method revisited. *Optimization Letters,* 2008, 2, 445-454.

[34] Hirsch, M.J.; Meneses, C.N.; Pardalos, P.M.; Resende, M.G.C. Global optimization by continuous GRASP. *Optimization Letters,* 2007, 1, 201-212.

[35] Hirsch, M.J.; Pardalos, P.M.; Resende, M.G.C. Speeding up continuous GRASP. *European Journal of Operational Research,* 2010, 205, 507-521.

[36] Meintjes, K.; Morgan, A.P. Chemical equilibrium systems as numerical test problems. *ACM Transactions on Mathematical Software,* 1990, 16, 143-151.

[37] Maranas, C.D.; Floudas, C.A. Finding all solutions of nonlinearly constrained systems of equations. *Journal of Global Optimization,* 1995, 7, 143-182.

[38] Van Hentenryck, P.; McAllester, D.; Kapur, D. Solving polynomial systems using a branch and prune approach. *SIAM Journal on Numerical Analysis,* 1997, 34, 797-827.

[39] Grosan, C.; Abraham, A. A new approach for solving nonlinear equations systems. *IEEE Transactions on Systems, Man, and Cybernetics A,* 2008, 38, 698-714.

[40] Sacco, W.F.; Alves Filho, H.; de Oliveira, C.R.E. A populational Particle Collision Algorithm Applied to a Nuclear Reactor Core Design Optimization. *Joint International Topical Meeting on Mathematics & Computation and Supercomputing in Nuclear Applications (M&C + SNA 2007),* American Nuclear Society, Monterey, 2007.

[41] Goldberg, D.E. *Genetic Algorithms in Search, Optimization and Machine Learning*. Addison-Wesley, Reading, MA, 1989.

[42] da Luz, E.F.P.; Becceneri, J.C.; Campos Velho, H.F. Multiple Particle Collision Algorithm applied to Radiative Transference and Pollutant Localization Inverse Problems. *2011 IEEE International Parallel & Distributed Processing Symposium*, 347-351, Anchorage, Alaska, 2011.

[43] Abuhamdah, A.; Ayob, M. Multi-neighbourhood particle collision algorithm for solving course timetabling problems. *Proceedings of the 2009 Second Conference on Data Mining and Optimization*, 21-27, IEEE, Selangor, Malaysia, 2009.

[44] Landa-Silva, D.; Obit, J.H. Great Deluge with Non-linear Decay Rate for Course Timetabling Problems. *2008 4th International IEEE Conference on Intelligent Systems,* 8-11-8-18, IEEE, Varna, Bulgaria, 2008.

[45] Luus, R.; Jaakola, T.H.I. Optimization by direct search and systematic reduction of the size of search region. *AIChE Journal,* 1973, 19, 760-766.

[46] Lee, Y.P.; Rangaiah, G.P.; Luus, R. Phase and chemical equilibrium calculations by direct search optimization. *Computers and Chemical Engineering,* 1999, 23, 1183-1191.

[47] Luus, R.; Hennessy, D. Optimization of fed-batch reactors by the Luus-Jaakola optimization procedure, *Industrial & Engineering Chemistry Research,* 1999, 38, 1948-1955.

[48] Luus, R. Use of Luus-Jaakola optimization procedure for singular optimal control problems, *Nonlinear Analysis*, 2001, 47, 5647–5658.

[49] Lei, J.; Liu, S.; Wang, X.Y. Generalised multi-scale image reconstruction algorithm for electrical capacitance tomography. *The Image Science Journal,* 2011, 49, 134-153.

[50] Floudas, C.A.; Pardalos, P.M. (Eds.). *Encyclopedia of Optimization*, 2nd Edition, 1272-1277, Springer, New York, 2009.

[51] Lee, K.S.; Geem, Z.W. A new meta-heuristic algorithm for continuous engineering optimization: harmony search theory and practice. *Computer Methods in Applied Mechanics and Engineering*, 2005, 194, 3902-3933.

In: Global Optimization
Editor: Angelika Michalski
ISBN: 978-1-62417-796-5

Chapter 5

CLASSIFIER-ASSISTED FRAMEWORKS FOR COMPUTATIONALLY EXPENSIVE OPTIMIZATION PROBLEMS

Yoel Tenne*
Ariel University Centre, Ariel, Israel

Abstract

The modern engineering design optimization process often replaces laboratory experiments with computer simulations. This setup formulates an optimization problem of a black-box function, namely, which has no analytic expression and which is computationally expensive to evaluate. However, computer simulations introduce an additional challenge into the design optimization process: often there will exist candidate designs which cause the simulation to fail, which implies that a large portion of the optimization resources may be wasted, and this can lead to a poor final result. To effectively handle such problems, this chapter describes two optimization frameworks which incorporate a classifier into the search. The classifier's role is to predict which solutions are expected to crash the simulation, and this prediction is then used to bias the search towards solutions for which the simulation is expected to succeed. A baseline framework is described which incorporates a single classifier, followed by a more elaborate framework which selects an optimal type of classifier out of a family candidates. Performance analysis using a representative simulation-driven engineering problem of airfoil shape optimization

*E-mail address: y.tenne@ariel.ac.il

shows the effectiveness of the proposed frameworks and highlights the merit of incorporating a classifier into the search.

PACS 05.45-a, 52.35.Mw, 96.50.Fm

Keywords: expensive optimization problems, metamodelling, classification.

1. Introduction

Nowadays engineers extensively rely on computer simulations to evaluate candidate designs during the process of product design. Such simulations are employed since they make the design process more economical, and allow to investigate scenarios which may be complicated to evaluate by real-world experiments. These computer simulations, which must be properly validated with laboratory experiments, transform the design process into an optimization problem having three distinct features (Tenne and Goh, 2010):

- The simulation acts as the objective function since it assigns to candidate designs their corresponding objective values. However, the simulation is often a legacy code or a commercial software whose inner workings are inaccessible to the user. This results in a *black-box function*, namely, which lacks an analytic expression, and therefore precludes the use of some optimizers.
- Each simulation run is *computationally expensive*, that is, it requires a lengthy run time, and this severely restricts the number of candidate designs which can be evaluated.
- Both the real-world physics being modelled, and the numerical simulation process, may result in an objective function having a complicated nonconvex landscape which exacerbates the optimization difficulty.

These scenarios are commonly referred to in the literature as *expensive black-box optimization problems*, and a variety of algorithms have been proposed to address them (Tenne and Goh, 2010; Forrester and Keane, 2008; Queipo et al., 2005).

On top of the above challenges, such problems often present another difficulty since for some candidate designs the simulation will fail and no objective value will be provided. In this study such designs are termed *simulator-infeasible* (SI), while those for which the simulation succeeds are termed *simulator-feasible* (SF). SI designs have two main implications on the optimization search:

- Since they do not have a corresponding objective value, the objective function becomes discontinuous, which introduces an additional optimization difficulty.

and

- Such designs can consume a large portion of the allotted computational resources yet they do not provide any objective values, and this can lead to search stagnation and to a poor final result.

In this study it is assumed that the simulation failures are deterministic, namely, evaluations of a SF design will consistently succeed, while those of a SI design will consistently fail. It is also assumed that it is unknown prior to the optimization process why the simulation fails. Such failures may be caused by various reasons such as an inability of the simulation to handle complex shape geometries, or glitches in the simulation software.

Overall, the underlying settings and core assumptions on which this paper is based are:

- The optimization problem involves an expensive black-box objective function.

- The landscape of this objective function may be nonconvex and may contain other complicated features which exacerbate the difficulty of the optimization process.

- The search space contains candidate designs for which the simulation fails and does not return an objective value. It is unknown prior to the optimization search why this occurs, or how these candidate designs are distributed in the search space.

Numerous papers, such as Büche et al. (2005); Jin et al. (2002), and Poloni et al. (2000), have described encountering SI designs in engineering optimization problems, which indicates that such designs are prevalent in practise. Existing strategies to handle them include discarding such designs altogether without using them to adapt the optimization search, or assigning them a penalized objective value and then incorporating them into the optimization search. However, both approaches have significant demerits, for example, they discard information which can be used to improve the optimization search, or they can lead the optimizer to converge to a poor final solution.

In these settings, this chapter describes several optimization frameworks which aim to effectively handle expensive problems which contain SI designs. A main aspect of these frameworks is that they incorporate a *classifier* into the optimization search, where the classifier's role is to predict if a candidate design is SI or not, prior to its evaluation. This prediction is then used to bias the search to designs predicted to be SF. The effectiveness of the proposed frameworks is demonstrated by applying them to a simulation-driven engineering application of airfoil shape optimization, which is representative of industrial problems. Extensive analysis shows the merit of the proposed frameworks, and highlights the contribution of proposed classifier-assisted optimization process.

The remainder of this chapter is organized as follows: Section 2 provides the pertinent background information, Section 3 describes a baseline framework for optimization, and and Section 4 describes a more elaborate framework which employs a classifier selection step. Lastly, Section 5 concludes this chapter.

2. Background

2.1. Expensive optimization problems and computational intelligence algorithms

Expensive optimization problems arise in diverse domains across engineering and science, and as mentioned in Section 1, a typical scenario is simulation-driven design optimization problems. Figure 1 shows the layout of such problems, in which the simulation is viewed as a black-box function, namely, it assigns objective values to candidate designs while its analytic expression is unknown. In this setup, candidate designs are represented as vectors of design variables and are provided as inputs to the simulation. The black-box objective function arising in this setup is often highly nonconvex (which exacerbates the optimization difficulty), where its nonconvexity stems from the underlying real-world phenomena being modelled or the numerical simulation process (Forrester and Keane, 2008; Prudius and Andradóttir, 2004).

A common strategy to address the challenge of computationally expensive evaluations is by generating a *metamodel* (also termed a *response surface* or a

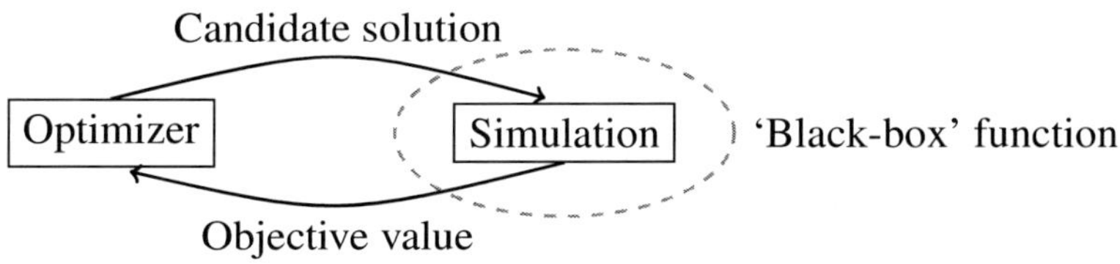

Figure 1. The layout of an expensive black-box optimization problem. The optimizer generates candidate solutions, and these are evaluated by the simulation to obtain their corresponding objective values. The optimizer views the simulation as a black-box function, that is, as having no analytic expression.

surrogate model) which provides the optimizer with predicted objective values at a much lower computational cost when compared to the simulation code. Metamodels are typically interpolants trained with previously evaluated vectors, and variants include artificial neural networks, Kriging, polynomials, and radial basis functions (RBF) (Forrester and Keane, 2008; Queipo et al., 2005).

After training a metamodel, typically a search is performed to locate its optimum, and two main classes of algorithms can be employed:

- Gradient-based optimizers: These optimizers typically operate on a single candidate solution which is refined at each iteration based on the local gradient. The resultant search is typically localized.

- Direct search optimizers: Such optimizers typically employ a number of candidate solutions concurrently, and manipulate them based only on the observed function values (gradients are not utilized). The resultant search tends to be more explorative when compared to gradient-based methods.

Since black-box functions often have a complicated nonconvex landscape, gradient-based optimizers can exhibit rapid convergence to a local optimum and therefore provide a poor final solution. This has motivated the use of direct-search optimizers in such problems since their explorative search behaviour often allows to them to detect a better optimum in such settings (Zahara and Kao, 2009; Babu and Rakesh, 2006). One such optimizer, which has been widely used in literature, and which is therefore employed in this study, is the *evolutionary algorithm* (EA). It employs the following nature-inspired operators (de Jong, 2006): i) selection (the candidate solutions with the best objective value are selected as *parents*), ii) recombination (parents are combined to produce offspring), and iii) mutation (a random change is applied to the components of some of the offspring). The offspring population is then evaluated, and the members with the best objective values serve as the next generation population. The process then repeats until a termination criterion is met, for example, if the maximum number of generations has been reached. Through these operators the EA emphasizes the best candidate solutions, which leads its population to explore the function landscape and to eventually cluster around an optimum. Algorithm 1 gives a pseudocode of a baseline EA.

Algorithm 1: A baseline evolutionary algorithm (EA)

```
initialize and evaluate a population of candidate solutions;
/* main loop                                                  */
repeat
    select a group of candidate solutions and designate them as parents;
    recombine the parents to create offspring;
    mutate some of the offspring;
    evaluate the offspring;
    select the candidate solutions which will comprise the population of
    the next generation;
until convergence or maximum search duration;
```

However, while metamodel-assisted frameworks address the issue of expensive evaluations, a main challenge when using metamodels is that of optimization with approximate (and inaccurate) objective values. Specifically, since the true objective function can be evaluated only a small number of times, only a small number of vectors will be available to train the metamodel, which in turn degrades its accuracy. This adversely affects the search and may even lead the optimizer to converge to a *false optimum*, namely, an optimum of the metamodel which is not an optimum of the true expensive function (Jin et al., 2002). Therefore, it is necessary to *manage* the metamodel to safeguard its accuracy. To accomplish this, the frameworks described in this chapter employ a *trust-region* (TR) approach, which originated in the field of nonlinear programming (Conn et al., 1997). The idea is to perform a sequence of *trial steps*, where each is constrained to the TR, namely, the region where the metamodel is assumed to be accurate. Based on the trial step success, that is, whether a better solution was indeed found, the TR and metamodel are updated, and the sequence repeats until some termination condition is met. A merit of the TR approach is that it ensures asymptotic convergence to an optimum of the true expensive function (Conn et al., 1997). Sections 3 and 4 give a detailed description of the TR approach implemented in the two frameworks discussed in this chapter.

2.2. Simulator-infeasible vectors

As mentioned in Section 1, this chap[ter focuses on expensive optimization problems which contain simulator-infeasible (SI) vectors, namely, which cause the simulation to fail. Numerous studies have acknowledged the existence of such vectors and the difficulties they introduce into the optimization search. For example, Poloni et al. (2000) mentioned observing "failure of the simulation code" during an engineering optimization problem, Booker et al. (1999) highlighted that during their simulation-driven optimization "attempts to evaluate the objective function failed", and similarly, Büche et al. (2005) noted that in some of their simulation runs "evaluation of all points fails". Additional examples include Koehler and Owen (1996); Conn et al. (1998) and Okabe (2007).

As evident from these references, SI vectors are common in industrial optimization problems, and therefore several techniques have been explored to handle them. For example, Rasheed et al. (1997) performed design optimization using an EA, and used a classifier to screen vectors before evaluating them with the simulation. Those predicted to be SI were assigned a 'death penalty', that is, a fictitious and highly penalized objective value, to quickly eliminate them from the population. The authors did not consider the use of metamodels, and in their implementation the EA evaluated the expensive function, namely, the simulation, directly. In another related study, Emmerich et al. (2002) also used the penalty approach, but incorporated the penalized vectors into the metamodel to bias the search towards SF solutions. In contrast, Büche et al. (2005) ignored the SI vectors altogether when training the metamodel, and used only the SF ones. These strategies, and similar ones, have several demerits in the context of expensive optimization problems:

- Assigning SI vectors a penalized objective value and then incorporating them into the metamodel can severely deform the metamodel landscape and degrade its prediction accuracy.
- Discarding SI vectors results in loss of information which could have been

used to improve the optimization search.

As an example, Figure 2 shows the effect of penalizing SI vectors and incorporating them into the metamodel. The left figure shows a metamodel which was trained using 30 SF vectors, while the right figure shows the resultant metamodel when 20 SI vectors were assigned a penalized objective value, taken as the worst objective value of the SF vectors, and incorporated into the metamodel. This resulted in a severe deformation of the metamodel landscape, and severely degraded its prediction accuracy.

Such issues have motivated exploring alternative approaches for handling SI vectors. For example, Tenne and Armfield (2008) proposed a dual metamodel approach, where one metamodel was used for the objective function, and another provided an interpolated penalty which was proportional to the distance from the SI vectors. Other studies used classifiers for constrained nonlinear programming, though unrelated to SI vectors (Handoko et al., 2010). Further exploring the use of classifiers, Tenne et al. (2010) obtained preliminary results with a classifier-assisted algorithm for handling SI vectors.

3. A baseline framework

This section describes in detail a baseline framework for expensive optimization problems which contain SI vectors (Tenne et al., 2011a). As mentioned in Section 1, a main concept of the proposed framework is that it incorporates a classifier into the optimization search, where the goal of the classifier is to predict if a candidate vector is SF or SI prior to its evaluation by the metamodel. This prediction is then used to bias the search towards vectors predicted to be SF.

3.1. Workflow

The framework begins by generating an initial sample of vectors based on a Latin hypercube (LH) design of experiments (McKay et al., 1979). The technique is used to ensure the sample is space-filling, as this improves the accuracy of the resultant metamodel. Briefly, for a sample of l vectors the range of each variable is split into l equal intervals, and one point is sampled at random in each interval. Next, a sample point is selected at random and without replacement for each variable, and these samples are combined to produce a vector. This pro-

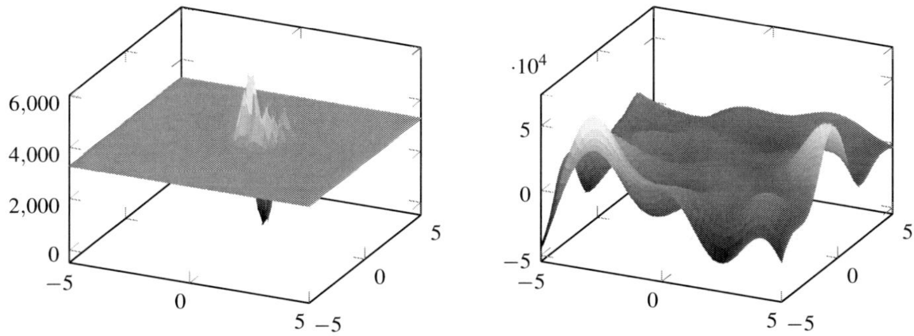

(a) (b)

Figure 2. An example showing the effect of incorporating penalized SI vectors into the metamodel. The objective function employed was Rosenbrock. (a) shows a Kriging metamodel trained using a sample of 30 SF vectors, and (b) shows the resultant metamodel when 20 SI vectors were added and assigned as a penalized objective value the worst objective value of the SF vectors. This severely deformed the metamodel landscape and degraded its prediction accuracy.

cedure is repeated for l times to generate the complete sample, which is then evaluated with the expensive simulation, and the vectors are stored in memory.

The main optimization loop then begins, in which a metamodel is trained by using only the SF vectors stored in memory, and then a classifier is trained based on *all* the vectors stored in memory, namely, both the SF and SI ones, to account for both vector classes. In this implementation, a k nearest neighbours (kNN) classifier and a Kriging metamodel were employed, which are described in Appendix A and Appendix B, respectively. Next, a TR trial step is performed: the best vector in the memory storage is taken as the TR centre ($\boldsymbol{x}_{\mathrm{b}}$), and the framework seeks the optimum of the metamodel in the TR. As the optimizer, the framework uses the real-coded EA of Chipperfield et al. (1994), and since obtaining predicted objective values from the metamodel is computationally cheap, the EA uses a large population and many generations to improve its search. Table 1 gives the complete parameter settings of the EA. During this trial step, the EA receives the predicted objective values not from the metamodel but from the following *modified objective function*

$$\hat{m}(\boldsymbol{x}) = \begin{cases} m(\boldsymbol{x}) & \text{if the classifier predicts } \boldsymbol{x} \text{ is SF} \\ p & \text{if the classifier predicts } \boldsymbol{x} \text{ is SI} \end{cases} \quad (1)$$

where $m(\boldsymbol{x})$ is the metamodel prediction, and p is a penalized objective value taken to be the worst function value from the initial LH sample. Accordingly, the EA receives the metamodel prediction if the classifier predicts a vector is SI, but it receives the penalized objective value otherwise. In this setup, the information about SI vectors is preserved in the classifier, but they are *not* incorporated into the metamodel with a penalized objective value and therefore do not deform its landscape, which avoids the issues discussed in Section 2.2.

The vector found by the EA ($\boldsymbol{x}^{\star}$) is then evaluated with the true expensive function to obtain its objective value $f(\boldsymbol{x}^{\star})$. Following the classical TR approach (Conn et al., 2000), based on the outcome of the trial step the framework now updates the metamodel and the TR, as follows:

- If $f(\boldsymbol{x}^{\star}) < f(\boldsymbol{x}_{\mathrm{b}})$: The trial step was successful since the predicted optimum is indeed better than the current best solution, namely, $\boldsymbol{x}_{\mathrm{b}}$. Accordingly, the TR is centred at the new optimum, and the TR radius is doubled.

- If $f(\boldsymbol{x}^{\star}) \geqslant f(\boldsymbol{x}_{\mathrm{b}})$ *and* there are sufficient SF vectors inside the TR: The search was unsuccessful since the predicted optimum is not better than the current best vector. However, since there are sufficient SF vectors in the TR, the metamodel is considered to be sufficiently accurate to justify contracting the TR, and therefore the TR radius is halved.

- If $f(\boldsymbol{x}^{\star}) \geqslant f(\boldsymbol{x}_{\mathrm{b}})$ *and* there are insufficient SF vectors inside the TR: The search was unsuccessful, but this may be due to a poor accuracy of the

Table 1. Internal parameters of the EA utilized in this study.

Population size	100
Generations	100
Selection	Stochastic universal selection (SUS)
Recombination	Intermediate, applied with probability $p = 0.7$
Mutation	Breeder Genetic Algorithm (BGA) mutation, applied with probability $p = 0.1$
elitism	10%

metamodel resulting from too few SF vectors in the TR. Therefore, the framework samples new vectors in the TR, as explained below.

Compared to the classical TR framework, the framework described contracts the TR only when there are sufficient vectors in the TR, to safeguard against premature convergence resulting from rapid contraction of the TR when the metamodel is inaccurate (Conn et al., 2000).

Another change from the classical TR approach is the sampling of a new vector ($\boldsymbol{x}_{\mathrm{n}}$) to improve the metamodel accuracy in the TR. To accomplish this, the new vector should be far from existing ones (Madych, 1992), which mathematically is formulated as

$$\boldsymbol{x}_{\mathrm{n}} : \max_{\boldsymbol{x} \in \mathcal{T}} \min_{\boldsymbol{x}_i \in \mathcal{T}} \{ \| \boldsymbol{x} - \boldsymbol{x}_i \|_2 \} \tag{2}$$

where $\boldsymbol{x}_i$, $i = 1 \ldots r$ are the existing, (namely, previously sampled) vectors which are in the TR (Johnson et al., 1990). To simplify the solution of (2) the proposed algorithm generates a LH sample in the TR and chooses the sample vector with the largest minimum distance.

To complete the description of the baseline framework, Algorithm 2 gives its pseudocode.

3.2. Performance analysis of the baseline framework

To evaluate the effectiveness of the proposed framework it was applied to a test problem of airfoil shape optimization, which is representative of industrial simulation-driven optimization problems, and whose formulation is as follows. During flight, an aircraft generates *lift*, which is the force which counters the aircraft weight and keeps it airborne, and *drag*, which is an aerodynamic friction force which obstructs the aircraft's movement. Both the lift and drag are strongly determined by the wing cross-section, namely, the *airfoil*. The optimization goal is then to find an airfoil shape which generates the lift and minimizes the drag. In practise, the design requirements for airfoils are often specified in terms of the nondimensional lift and drag coefficients, c_l and c_d, respectively, defined as

$$c_l = \frac{L}{\frac{1}{2}\rho V^2 S} \tag{3a}$$

$$c_d = \frac{D}{\frac{1}{2}\rho V^2 S} \tag{3b}$$

Algorithm 2: Proposed baseline framework for classifier-assisted optimization

```
generate an initial LH sample;
evaluate the sample vectors with the true expensive function and store
them in memory;
repeat
    /* metamodel and classifier training                                 */
    train a metamodel using the SF vectors stored in memory;
    train a classifier using all the vectors stored in memory;
    /* TR trial step                                                     */
    designate the best vector stored in memory as the TR centre;
    search for an optimum of the metamodel in the TR based on the
    modified objective function (1);
    evaluate the optimum found with the true expensive objective
    function;
    /* updates based on the trial step                                   */
    if the predicted optimum is better than the TR centre then
        double the TR radius
    else if the predicted optimum is not better than the TR centre and
    there is an insufficient number of vectors in the TR then
        add a new vector in the TR to improve the metamodel accuracy;
    else if the predicted optimum is not better than the TR centre and
    there is a sufficient number of vectors in the TR then
        halve the TR radius;
    add to the memory storage all the new vectors evaluated with the true
    expensive function;
until maximum number of analyses reached;
```

where L and D are the lift and drag forces, respectively, ρ is the air density, V is the aircraft speed, and S is a reference area, such as the wing area. Besides the lift and drag coefficients, the other relevant physical quantities involved are the aircraft altitude, speed, and *angle of attack* (AOA), which is the angle between the aircraft velocity and the airfoil *chord line*. Figure 3 shows a schematic diagram of the airfoil problem.

Airfoils were represented with the Parametric Sections (PARSEC) parametrization (Sobieszczansk-Sobieski and Haftka, 1997) which defines 11 design variables representing geometrical features, as shown in Figure 3. To ensure a closed airfoil shape the setting $dz_{TE} = 0$ was enforced, while bounds on the other variables were set based on Tenne et al. (2010), and are given in Table 2. The lift and drag of candidate airfoils were obtained from *XFoil*, a computational fluid dynamics simulation for analysis of subsonic isolated airfoils (Drela and Youngren, 2001). Each airfoil evaluation required up to 30 seconds on a desktop computer.

Several AOA settings were employed (2°,5°,10°and 15°), to evaluate the effectiveness of the framework in different problem settings. In particular, higher AOA values result in more frequent simulation failures since the corresponding flow field being modelled is more turbulent and is therefore more difficult to simulate numerically. Two variants of the framework described were employed:

- P-SVM: A variant which employed a support vector machine (SVM) classifier with Gaussian kernels, and
- P-KNN: A variant which employed a kNN classifier ($k = 3$) .

Details of these classifiers are given in Appendix A. Also included in the tests

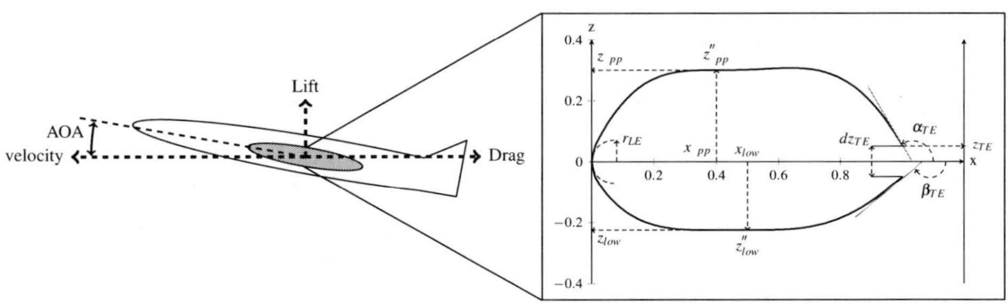

Figure 3. Physical quantities and airfoil parametrization in the airfoil problem employed with the baseline framework.

Table 2. Bounds on the PARSEC variables

Variable	Geometric feature	L.bound	U.bound
r_{LE}	leading-edge radius	0.001	0.003
x_{upp}	max. upper thickness location	0.175	0.5
z_{upp}	max. upper thickness	0	0.2
z''_{upp}	max. upper curvature	-2	-0.05
x_{low}	max. lower thickness location	0.175	0.5
z_{low}	max. upper thickness	-0.2	0
z''_{low}	max. lower curvature	0.05	2
z_{TE}	trailing edge height	-0.1	0.1
dz_{TE}	trailing edge thickness	0	0
$\alpha^{[1]}$	upper trailing edge angle (deg.)	150	180
$\beta^{[1,2]}$	lower trailing edge angle (deg.)	150	210

[1] measured anti-clockwise from the x-axis.
[2] $\beta \geqslant \alpha$ to avoid intersecting curves.

were two penalty-based variants (termed the *reference algorithms*) which use the same optimization steps as the baseline framework, *except* that they did not use a classifier but instead assigned the SI vectors a penalized objective value and incorporated them into the metamodel. The two variants used were:

- R-1: A reference algorithm with the penalized objective value taken to be the worst objective value from the initial Latin hypercube design (LHD) sample.
- R-10: As above, but the penalized objective value was taken to be 10 times the worst objective value in the initial sample.

This test setup was employed since it highlights the effect of the proposed classifier-assisted approach. In all tests the optimization limit was 200 function evaluations (simulation runs), and for valid statistical analysis tests were repeated for 30 times with each combination of algorithm and AOA.

Table 3 shows the statistics for the best objective function with the best mean and median emphasized at each AOA. It follows that the proposed framework with an SVM classifier (P-SVM) had the best mean in all cases and best median in two cases. The kNN variant had the best median for AOA=10°. Overall results show that the proposed framework outperformed the reference (penalty-based) algorithms, which also highlights the demerits of the penalty-based approach, namely: a) incorporating penalized vectors into the metamodel can deform its landscape and can therefore lead to an inferior final result, and b) performance can be sensitive to the penalty value but an optimal penalty value is unknown a-priori and it may be difficult to accurately estimate given the limited number of evaluations of the true expensive function.

Statistical significance analysis (at the $\alpha = 0.05$ level) shows that the performance gains of the P-SVM variant were not statistically significant at AOA=2° since at a low AOA settings the SI vectors were less prevalent and so the proposed framework operated during most of the optimization search similarly to the reference algorithms. At higher AOAs the SI vectors were more prevalent and there the results changed: at AOA=5° gains were significant over the R-10 variant and at AOA=10° they were significant over the R-1 and R-10 variants. At AOA=15° gains were borderline significant over the R-1 and significant over the R-10 variant.

4. An enhanced framework

The effectiveness of the framework described in the previous section depends on the type of classifier being used, but the optimal classifier type is problem dependant. In practice it may be impractical to identify an optimal type by trial optimization runs due to the limited number of evaluations of the true expensive function. Furthermore, using a single type of classifier, which may be unsuitable to the problem being solved, can degrade the search effectiveness. In these settings, this section describes an enhanced framework which leverages on that from the previous section, but which introduces a classifier selection stage in which the framework autonomously selects an optimal classifier out of a family of candidates (Tenne et al., 2011b).

4.1. Workflow

The enhanced framework follows the same steps as that of the baseline framework described in the previous section, except for the added classifier selection step. To accomplish this selection in a mathematically rigorous way, the framework leverages on the statistical *model-selection* theory (here a *model* refers to a functional relation which is used to relate observed inputs to outputs) and selects a classifier based on its estimated accuracy. The procedure, termed

Table 3. Statistics for the final objective value

AOA (deg)		P-SVM	P-KNN	R-1	R-10
	Mean	-2.851e+02	-9.981e+01	-9.637e+01	-2.550e+02
	SD	5.408e+02	3.179e+01	2.660e+01	8.370e+02
2	Median	-9.806e+01	-9.869e+01	-9.768e+01	-1.126e+02
	Min(best)	-2.375e+03	-1.835e+02	-1.670e+02	-4.604e+03
	Max(worst)	-3.181e+01	-1.084e+01	-3.425e+01	-3.070e+01
	Mean	-1.600e+03	-1.496e+03	-2.820e+02	-3.104e+02
	SD	7.392e+03	4.761e+03	3.973e+02	5.424e+02
5	Median	-8.810e+01	-1.016e+02	-9.320e+01	-8.405e+01
	Min(best)	-3.999e+04	-2.488e+04	-1.515e+03	-2.210e+03
	Max(worst)	-3.595e+01	-1.879e+01	-3.635e+01	-1.518e+01
	Mean	-2.416e+01	-2.293e+01	-1.697e+01	-2.103e+01
	SD	1.969e+01	1.198e+01	9.226e+00	1.102e+01
10	Median	-2.017e+01	-2.090e+01	-1.478e+01	-1.636e+01
	Min(best)	-1.216e+02	-7.533e+01	-4.130e+01	-4.233e+01
	Max(worst)	-1.237e+01	-8.926e+00	-6.545e+00	-4.784e+00
	Mean	-6.079e+00	-5.172e+00	-5.690e+00	-6.043e+00
	SD	2.038e+00	1.497e+00	2.111e+00	3.200e+00
15	Median	-5.322e+00	-5.321e+00	-5.009e+00	-4.598e+00
	Min(best)	-9.923e+00	-8.834e+00	-1.119e+01	-1.701e+01
	Max(worst)	-3.441e+00	-2.720e+00	-3.272e+00	-3.183e+00

P-SVM: proposed approach, SVM classifier.
P-*k*NN: proposed approach, *k*NN classifier.
R-1: reference approach, penalty = worst objective.
R-10: reference approach, penalty = 10× worst objective.

cross-validation (CV), operates as follows. The vectors which have been evaluated with the expensive simulation, and are therefore stored in memory, are split into a *training set* and a *testing set* in a 80–20 ratio. A candidate classifier is trained using the former set and its prediction accuracy is evaluated based on the latter set, and is calculated as

$$e = \sum_{i=1}^{l} \left(\hat{c}(\boldsymbol{x}_i) \neq F(\boldsymbol{x}_i) \right) , \qquad (4)$$

where $\hat{c}$ is the prediction of the trained classifier, $\boldsymbol{x}_i$, $i = 1 \ldots l$, are the CV testing vectors, and $F(\boldsymbol{x}_i)$ is the true and known class of the latter vectors, where in this study $F(\boldsymbol{x}_i) = 1$ was used for a SF vector, and $F(\boldsymbol{x}_i) = -1$ for a SI one. These class labels were used as they are common in literature, for example Wu et al. (2008). In the implementation described here, three candidate classifiers were well-established considered: kNN, linear discriminant analysis (LDA), and SVM, whose details are given in Appendix A. After selecting the best performing classifier type, the framework trains a new classifier by using all the vectors stored in memory, and this classifier is then used during the TR trial step.

In this setup, the optimal classifier type is re-examined with each iteration, which can result in frequent updating of the classifier type being employed during the optimization process. All others steps of this framework are identical to those of the baseline framework described in the previous section, and for completeness Algorithm 3 gives the pseudocode of the enhanced framework.

4.2. Performance analysis of the enhanced framework

Similarly to Section 3.2, the enhanced framework was evaluated by applying it to the problem of airfoil shape optimization. However, in this instance airfoils were represented with the Hicks-Henne parametrization (Hicks and Henne, 1978), which uses a baseline airfoil and adds the basis functions

$$b_i(x) = \left[\sin\left(\pi x^{\frac{\log(0.5)}{\log(i/(h+1))}}\right)\right]^4, \qquad (5)$$

with $i = 1 \ldots h$, where h is user-prescribed, to smoothly modify the baseline shape (Wu et al., 2003). The lower and upper curves of a candidate airfoil are then given by

$$y = y_b + \sum_{i=1}^{h} \alpha_i b_i(x), \qquad (6)$$

where y_b is the baseline upper/lower curve, which was taken as the NACA0012 symmetric airfoil, and $\alpha_i \in [-0.01, 0.01]$ are the coefficients (design variables) to be found. In this evaluation, $h = 10$ functions were used for the upper and lower curve, respectively, or a total of 20 coefficients, namely, design variables, per airfoil. Figure 4 gives the layout of the Hicks-Henne parametrization. As before, candidate airfoils were evaluated with the XFoil analysis code.

For a comprehensive evaluation, the proposed algorithm was also benchmarked against the following two representative metamodel-assisted algorithms:

Algorithm 3: Proposed enhanced framework for classifier-assisted optimization

```
generate an initial LH sample;
evaluate the sample vectors with the true expensive function and store
them in memory;
repeat
    /* metamodel training                                              */
    train a metamodel using the SF vectors stored in memory;
    /* classifier selection                                            */
    for candidate classifier = {kNN, LDA, SVM} do
        calculate the respective classification error (4);
    select the classifier with the lowest error and train a new classifier by
    using all the vectors stored in memory;
    /* TR trial step                                                   */
    designate the best vector stored in memory as the TR centre;
    search for an optimum of the metamodel in the TR based on the
    modified objective function (1);
    evaluate the optimum found with the true expensive objective
    function;
    /* updates based on the trial step                                 */
    if the predicted optimum is better than the TR centre then
        double the TR radius
    else if the predicted optimum is not better than the TR centre and
    there is an insufficient number of vectors in the TR then
        add a new vector in the TR to improve the metamodel accuracy;
    else if the predicted optimum is not better than the TR centre and
    there is a sufficient number of vectors in the TR then
        halve the TR radius;
    add to the memory storage all the new vectors evaluated with the true
    expensive function;
until maximum number of analyses reached;
```

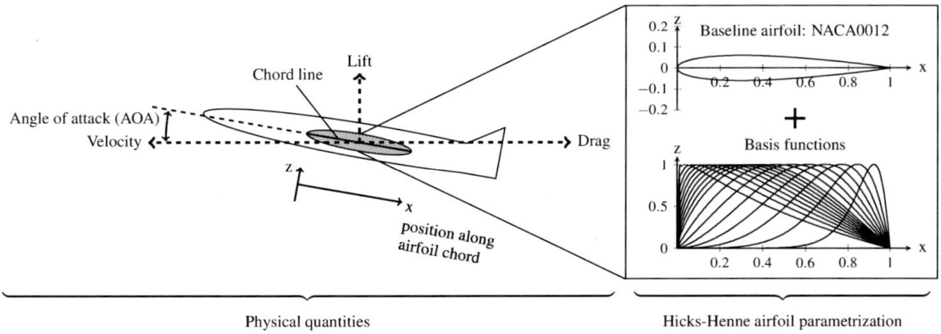

Figure 4. Physical quantities and airfoil parametrization in the airfoil problem (enhanced framework).

- *Metamodel assisted EA with periodic sampling* (EA–PS) (Ratle, 1999): The algorithm safeguards the metamodel accuracy by periodically evaluating a small subset of the population with the true objective function, and it then uses the latter to update the metamodel. The algorithm begins by generating and initial sample of vectors and it evaluates them with the expensive function. The main optimization loop then begins, where the algorithm trains a metamodel by using the vectors evaluated so far, and uses a real-coded EA to search for the optimum of the metamodel. After the search is completed, the algorithm evaluates the best ten candidate solutions in the population with the true expensive function. The entire process then repeats, until the maximum number of expensive function evaluations is reached. Following Ratle (1999), the algorithm used the Kriging metamodel, which is described in Appendix B.

- *Expected-Improvement with a metamodel-assisted CMA-ES* (EI–CMA-ES) (Büche et al., 2005): The algorithm combines a covariance matrix adaptation evolutionary strategy (CMA-ES) optimizer (Hansen and Ostermeier, 2001) with a Kriging metamodel, and updates the latter based on the expected improvement framework (Jones et al., 1998). The algorithm begins by generating an initial sample of vectors, and evaluates them with the true function. Its main loop then begins, where at each generation it trains a local Kriging metamodel based on the recently evaluated vectors, and the vectors stored in memory which are nearest to the best solution. A CMA-ES algorithm then searches for an optimum of the metamodel in the region enclosing the vectors which were used to train the metamodel. In the spirit of the expected improvement framework (Jones et al., 1998), the optimizer operates on the merit function

$$\hat{f}(\boldsymbol{x}) = m(\boldsymbol{x}) - \rho\zeta(\boldsymbol{x}), \tag{7}$$

where $m(\boldsymbol{x})$ is the Kriging metamodel prediction, ρ is a prescribed coefficient, and $\zeta(\boldsymbol{x})$ is the estimated Kriging prediction error, which is zero at sampled points since there the true objective value is known. The search is repeated for $\rho = 0, 1, 2,$ and 4, to obtain four solutions corresponding to different search profiles, namely, ranging from a local search ($\rho = 0$) to a more explorative one ($\rho = 4$). All non-duplicate solutions found are evaluated with the true expensive function and are stored in memory. In case no new solutions were evaluated, for example, because they already match solutions stored in memory, the algorithm generates a new solution by perturbing the current best one. Following Reference (Büche et al., 2005), the algorithm used a training set of 100 vectors comprising of the 50 most recently evaluated ones and the 50 nearest-neighbours, and the CMA-ES used the default values given in Hansen and Ostermeier (2001).

The tests also included a variant of the enhanced framework but with a fixed classifier type, namely, only *k*NN, to study the contribution of the classifier selection step. The variant is designated as ***KK*** (Kriging–*k*NN).

Tests were performed in three AOA settings (20°, 30°, and 40°). In all tests the optimization budget was 200 evaluations of the true expensive function (simulation runs), and the size of the initial sample was 20. To support a valid statistical analysis, 30 trials were repeated for each algorithm and test case combination.

Table 4 gives the test statistics for the three AOA cases, as well as the significance-level (α) at which the enhanced framework was better than each of the other algorithms, namely, EA–PS, EI–CMA-ES, and KK, where an empty entry indicates no statistically significant difference up to the 0.05 level. Statistical significance was measured by using the Mann–Whitney nonparametric test. For each AOA case, the best mean and median results are emphasized. From studying the test results for each AOA setting it follows:

- AOA=20°: The enhanced framework obtained the best mean score, and its performance gains over the KK and EA–PS variants were statistically significant at the $\alpha = 0.01$ level. The EI–CMA-ES algorithm had the best median result, followed by the enhanced framework.

- AOA=30°: The KK variant obtained the best mean and median, followed by the enhanced framework. The performance gains of the enhanced framework over those of the EI–CMA-ES algorithm were statistically significant at 0.01 level.

- AOA=40°: The enhanced framework achieved the best mean, and the EA–PS algorithm achieved the best median, closely followed by the enhanced framework. The performance gains of the enhanced framework over those of the EI–CMA-ES algorithm were statistically significant at the 0.01 level.

Overall, test results show that the enhanced framework performed well, as it consistently obtained either the best or near-best mean and median statistics. Results also highlight the contribution of the classifier selection stage, as in two cases (AOA = 20° and 40°) the enhanced framework outperformed the KK variant which did not select a classifier during the optimization search.

5. Conclusion

Modern engineering design extensively relies on computer simulations, a scenario which formulates the problem of optimizing a computationally expensive black-box function. Often, such problems will contain candidate designs which cause the simulation to fail before completing the analysis, and therefore no output value will be provided to the optimizer. This, in turn, can degrade the search effectiveness and lead to a poor final result. To effectively handle this issue, this chapter has described two optimization frameworks which incorporate a classifier into the optimization search. The baseline framework employs a single type of classifier, while an enhanced framework autonomously selects the best performing classifier from a prescribed family of candidates. In both frameworks, the classifier is used to predict if a candidate design will cause the simulation to fail, and this prediction is then used to bias the search towards designs for which the simulation is expected to succeed. Performance analysis using a simulation-driven problem of airfoil shape optimization, which is representative of industrial applications, showed the effectiveness of the proposed frameworks and their performance gains over several existing approaches.

Table 4. Statistics for the final objective value

AOA		P	KK	EA–PS	EI–CMA-ES
20°	Mean	-1.035e+01	-8.091e+00	-6.889e+00	-1.023e+01
	SD	1.326e+00	1.697e+00	6.526e-01	2.025e+00
	Median	-1.049e+01	-7.283e+00	-6.843e+00	-1.107e+01
	Min(best)	-1.302e+01	-1.138e+01	-8.837e+00	-1.192e+01
	Max(worst)	-7.143e+00	-5.880e+00	-5.794e+00	-5.442e+00
	α		0.01	0.01	
30°	Mean	-3.155e+00	-3.192e+00	-3.146e+00	-2.910e+00
	SD	4.694e-02	3.105e-02	3.345e-02	4.761e-02
	Median	-3.140e+00	-3.183e+00	-3.140e+00	-2.916e+00
	Min(best)	-3.270e+00	-3.298e+00	-3.223e+00	-3.005e+00
	Max(worst)	-3.091e+00	-3.145e+00	-3.092e+00	-2.813e+00
	α				0.01
40°	Mean	-2.793e+00	-2.784e+00	-2.784e+00	-2.552e+00
	SD	3.380e-02	2.230e-02	4.732e-02	4.249e-02
	Median	-2.785e+00	-2.782e+00	-2.786e+00	-2.557e+00
	Min(best)	-2.875e+00	-2.827e+00	-2.869e+00	-2.637e+00
	Max(worst)	-2.726e+00	-2.742e+00	-2.717e+00	-2.455e+00
	α				0.01

P: Proposed enhanced framework.
KK: Proposed enhanced framework restricted to a Kriging metamodel and a *k*NN classifier.
EA–PS: EA with periodical sampling (Ratle, 1999).
EI–CMA-ES: Expected Improvement framework with a CMA-ES optimizer (Büche et al., 2005).

A. Classifiers

Classifiers originated in the domain on machine learning, and their goal is predict the class of new vectors, that is, given the vectors $\boldsymbol{x}_i \in \mathbb{R}^d$, $i = 1 \ldots n$, and their respective class labels $F(\boldsymbol{x}_i) \in \mathbb{I}$, for example, $\mathbb{I} = \{-1, +1\}$, a classifier performs the mapping

$$c(\boldsymbol{x}) : \mathbb{R}^d \to \mathbb{I}, \tag{8}$$

where $c(\boldsymbol{x})$ is the class assigned by the classifier.

Details of the classifiers considered in this chapter are as follows:

- *Linear Discriminant Analysis* (LDA): In a two-class problem, where the class labels are $F(\boldsymbol{x}_i) \in \mathbb{I} = \{-1, +1\}$, the classifier attempts to model the conditional probability density functions of a vector belonging to each class, where the latter functions are assumed to be normally distributed. The classifier considers the separation between classes as the ratio of: a) the variance between classes, and b) the variance within the classes , and obtains a vector $\boldsymbol{w}$ which maximizes this ratio. The vector $\boldsymbol{w}$ is such that it is orthogonal to the hyperplane separating the two classes. A new vector $\boldsymbol{x}$ is classified based on its projection with respect to the separating hyperplane, that is,

$$c(\boldsymbol{x}) = \operatorname{sign}(\boldsymbol{w} \cdot \boldsymbol{x}). \tag{9}$$

- *k Nearest Neighbours* (KNN): The classifier assigns the new vector the same class as its closest training vector, namely:

$$c(\boldsymbol{x}) = F(\boldsymbol{x}_{\text{NN}}) : d(\boldsymbol{x}, \boldsymbol{x}_{\text{NN}}) = \min d\big(\boldsymbol{x}, \boldsymbol{x}_i\big),\ i = 1 \ldots n, \tag{10}$$

where $d(\boldsymbol{x}, \boldsymbol{y})$ is distance measure such as the l_2 norm. An extension of this technique is to assign the class most frequent among the $k > 1$ nearest neighbours (kNN). In this study the classifier used $k = 3$.

- *Support Vector Machines* (SVM): The classifier projects the data into a high-dimensional space where it can be more easily separated into disjoint classes. In a two-class problem, and assuming class labels $F(\boldsymbol{x}_i) \in \mathbb{I} = \{-1, +1\}$, an SVM classifier tries to find the best classification function for the training data. For a linearly separable training set, a linear classification function is the separating hyperplane passing through the middle of the two classes. Once this hyperplane has been fixed, new vectors are classified based on their relative position to this hyperplane, that is, whether they are "above" or ""below" it. Since there are many possible separating hyperplanes, an SVM classifier adds the condition that the hyperplane should maximize its distance to the its nearest vectors from each class. This is accomplished by maximizing the Lagrangian

$$L_P = \frac{1}{2}\|\boldsymbol{w}\| - \sum_{i=1}^{n} \alpha_i F(\boldsymbol{x}_i)(\boldsymbol{w} \cdot \boldsymbol{x}_i + b) + \sum_{i=1}^{n} \alpha_i, \tag{11}$$

where n is the number of samples (training vectors), $F(\boldsymbol{x}_i)$ is the class of the ith training vector, and $\alpha_i \geqslant 0$, $i = 1 \ldots n$, are the Lagrange multipliers, such that the derivatives of L_P with respect to α_i are zero. The vector $\boldsymbol{w}$ and the scalar b define the hyperplane.

B. Kriging metamodel

The Kriging metamodel takes a statistical approach to interpolation by combining two components: a 'drift' function, which is a global coarse approximation of the true function, and a local correction based on the correlation between the interpolation vectors. Given a set of evaluated vectors, $\boldsymbol{x}_i \in \mathbb{R}^d$, $i = 1 \ldots n$, the metamodel is trained such that it exactly interpolates the observed values, that is, $m(\boldsymbol{x}_i) = f(\boldsymbol{x}_i)$, where $m(\boldsymbol{x})$ and $f(\boldsymbol{x})$ are the metamodel and true objective function, respectively. Using a constant drift function gives the Kriging metamodel

$$m(\boldsymbol{x}) = \beta + \kappa(\boldsymbol{x}) , \tag{12}$$

with the drift function β and local correction $\kappa(\boldsymbol{x})$. The latter is defined by a stationary Gaussian process with mean zero and covariance

$$Cov[\kappa(\boldsymbol{x})\kappa(\boldsymbol{y})] = \sigma^2 c(\theta, \boldsymbol{x}, \boldsymbol{y}) , \tag{13}$$

where $c(\theta, \boldsymbol{x}, \boldsymbol{y})$ is a user-prescribed correlation function. A common choice for the latter is the Gaussian correlation function (Forrester and Keane, 2008), defined as

$$c(\theta, \boldsymbol{x}, \boldsymbol{y}) = \prod_{i=1}^{d} \exp\left(-\theta\,(x_i - y_i)^2\right) , \tag{14}$$

and combining it with the constant drift function transforms the metamodel from (12) into the following form

$$m(\boldsymbol{x}) = \hat{\beta} + \boldsymbol{r}(\boldsymbol{x})^{\mathrm{T}} \boldsymbol{R}^{-1} (\boldsymbol{f} - \mathbf{1}\hat{\beta}) . \tag{15}$$

Here, $\hat{\beta}$ is the estimated drift coefficient, $\boldsymbol{R}$ is the symmetric matrix of correlations between all interpolation vectors, $\boldsymbol{f}$ is the vector of objective values, and $\mathbf{1}$ is a vector with all elements equal to 1. $\boldsymbol{r}^{\mathrm{T}}$ is the correlation vector between a new vector $\boldsymbol{x}$ and the sample vectors, namely,

$$\boldsymbol{r}^{\mathrm{T}} = [c(\theta, \boldsymbol{x}, \boldsymbol{x}_1), \ldots, c(\theta, \boldsymbol{x}, \boldsymbol{x}_n)] . \tag{16}$$

The estimated drift coefficient $\hat{\beta}$ and variance $\hat{\sigma}^2$, which are required in Equation (15), are obtained as follows

$$\hat{\beta} = \left(\mathbf{1}^{\mathrm{T}} \boldsymbol{R}^{-1} \mathbf{1}\right)^{-1} \mathbf{1}^{\mathrm{T}} \boldsymbol{R}^{-1} \boldsymbol{f} , \tag{17a}$$

$$\hat{\sigma}^2 = \frac{1}{n} \left[(\boldsymbol{f} - \mathbf{1}\hat{\beta})^{\mathrm{T}} \boldsymbol{R}^{-1} (\boldsymbol{f} - \mathbf{1}\hat{\beta}) \right] . \tag{17b}$$

Fully defining the metamodel requires the correlation parameters $\boldsymbol{\theta}$, which are commonly taken as the maximizers of the metamodel likelihood. This is achieved by minimizing the expression (Sacks et al., 1989)

$$\psi(\boldsymbol{\theta}) = |\boldsymbol{R}|^{1/n} \hat{\sigma}^2 , \tag{18}$$

which is a function only of $\boldsymbol{\theta}$ and the sample data. The formulation given above fully defines the Kriging metamodel (Queipo et al., 2005; Forrester and Keane, 2008; Sacks et al., 1989).

While a different correlation parameter can be used for each variable, the metamodel used in this study, as is commonly done in the literature, is *isotropic*, namely, it employs a single correlation parameter. This results in a univariate likelihood function, which is easier to optimize (Martin and Simpson, 2005). The minimization of (18) was achieved with the numerical code of Lophaven et al. (2002) which employs the Hooke-Jeeves pattern search algorithm (Hooke and Jeeves, 1961), which is a gradient-free direct search optimizer. Numerically, in this study the steps involved in generating the Kriging metamodel were as follows:

Step 1) The sample data, namely, the vectors and their corresponding objective values, were normalized to have a mean zero and standard deviation of one. This was done only for internal representation, and the final results were then converted back into the original coordinates.

Step 2) Internal parameters, such as the matrix of correlations between sample sites, were calculated.

Step 3) The optimal correlation parameter (θ^*) was obtained by minimizing $\psi(\theta)$ from equation (18) using the Hooke-Jeeves direct search algorithm.

Using an isotropic Kriging metamodel simplified the optimization of $\psi(\theta)$, and no restarts of the direct search algorithm were needed.

References

Y. Tenne, C. K. Goh (Eds.), Computational Intelligence in Expensive Optimization Problems, vol. 2 of *Evolutionary Learning and Optimization*, Springer, 2010.

A. I. J. Forrester, A. J. Keane, Recent advances in surrogate-based optimization, Progress in Aerospace Science 45 (1–3) (2008) 50–79.

N. V. Queipo, R. T. Haftka, W. Shyy, T. Goel, R. Vaidyanathan, K. P. Tucker, Surrogate-based analysis and optimization, Progress in Aerospace Science 41 (2005) 1–28.

D. Büche, N. N. Schraudolph, P. Koumoutsakos, Accelerating Evolutionary Algorithms With Gaussian Process Fitness Function Models, IEEE Transactions on Systems, Man, and Cybernetics–Part C 35 (2) (2005) 183–194.

Y. Jin, M. Olhofer, B. Sendhoff, A framework for evolutionary optimization with approximate fitness functions, IEEE Transactions on evolutionary computation 6 (5) (2002) 481–494.

C. Poloni, A. Giurgevich, L. Onseti, V. Pediroda, Hybridization of a Multi-objective Genetic Algorithm, a Neural Network and a Classical Optimizer for a Complex Design Problem in Fluid Dynamics, Computer Methods in Applied Mechanics and Engineering 186 (2–4) (2000) 403–420.

A. E. Prudius, S. Andradóttir, Simulation Optimization using Balanced Explorative and Exploitative Search, in: R. G. Ingalls (Ed.), Proceedings of the 2004 Winter Simulation Conference–WSC 2004, IEEE, New York, 545–549, 2004.

E. Zahara, Y.-T. Kao, Hybrid Nelder-Mead simplex search and particle swarm optimization for constrained engineering design problems, Expert Systems with Applications 36 (2009) 3880–3886.

B. V. Babu, A. Rakesh, Modified Differential Evolution (MDE) for Optimization of Non-Linear Chemical Processes, Computers and Chemical Engineering 30 (6–7) (2006) 989–1002.

K. A. de Jong, Evolutionary Computation:A Unified Approach, MIT Press, 2006.

A. R. Conn, K. Scheinberg, P. L. Toint, On the Convergence of Derivative-Free Methods for Unconstrained Optimization, in: A. Iserles, M. D. Buhmann (Eds.), Approximation Theory and Optimization: Tributes to M.J.D. Powell, Cambridge University Press, Cambridge; New York, 83–108, 1997.

A. J. Booker, J. E. Dennis, P. D. Frank, D. B. Serafini, V. Torczon, M. W. Trosset, A Rigorous Framework for Optimization of Expensive Functions by Surrogates, Structural Optimization 17 (1) (1999) 1–13.

J. R. Koehler, A. B. Owen, Computer Experiments, in: S. Ghosh, C. R. Rao, P. R. Krishnaiah (Eds.), Handbook of Statistics, Elsevier, Amsterdam, 261–308, 1996.

A. R. Conn, K. Scheinberg, P. L. Toint, A Derivative Free Optimization Algorithm in Practice, in: Proceedings of the Seventh AIAA/USAF/NASA/ISSMO Symposium on Multidisciplinary Analysis and Optimization, American Institute of Aeronautics and Astronautics, Reston, Virginia, AIAA paper number AIAA-1998-4718, 1998.

T. Okabe, Stabilizing Parallel Computation for Evolutionary Algorithms on Real-World Applications, in: Proceedings of the *7th* International Conference on Optimization Techniques and Applications–ICOTA 7, Universal Academy Press, Tokyo, 131–132, 2007.

K. Rasheed, H. Hirsh, A. Gelsey, A Genetic Algorithm for Continuous Design Space Search, Artificial Intelligence in Engineering 11 (1997) 295–305.

M. T. M. Emmerich, A. Giotis, M. Özedmir, T. Bäck, K. C. Giannakoglou, Metamodel-Assisted Evolution Strategies, in: J. J. Merelo Guervós (Ed.), The *7th* International Conference on Parallel Problem Solving from Nature–PPSN VII, no. 2439 in Lecture Notes in Computer Science, Springer, Berlin, 361–370, 2002.

Y. Tenne, S. W. Armfield, A Versatile Surrogate-Assisted Memetic Algorithm for Optimization of Computationally Expensive Functions and its Engineering Applications, in: A. Yang, Y. Shan, L. Thu Bui (Eds.), Success in Evolutionary Computation, vol. 92 of *Studies in Computational Intelligence*, Springer-Verlag, Berlin; Heidelberg, 43–72, 2008.

S. Handoko, C. K. Kwoh, Y.-S. Ong, Feasibility Structure Modeling: An Effective Chaperon for Constrained Memetic Algorithms, IEEE Transactions on Evolutionary Computation 14 (5) (2010) 740–758.

Y. Tenne, K. Izui, S. Nishiwaki, Handling Undefined Vectors in Expensive Optimization Problems, in: C. Di Chio (Ed.), Proceedings of the 2010 EvoStar Conference, vol. 6024/2010 of *Lecture Notes in Computer Science*, Springer, Berlin, 582–591, 2010.

Y. Tenne, K. Izui, S. Nishiwaki, A Classifier-assisted Framework for Expensive Optimization Problems: A Knowledge-Mining Approach, in: Proceedings of the 2011 Learning and Intelligent Optimization Conference–LION 5, vol. 6683 of *Lecture Notes in Computer Science*, Springer, 161–175, 2011a.

M. D. McKay, R. J. Beckman, W. J. Conover, A comparison of three methods for selecting values of input variables in the analysis of output from a computer code, Technometrics 21 (2) (1979) 239–245.

A. Chipperfield, P. Fleming, H. Pohlheim, C. Fonseca, Genetic Algorithm TOOLBOX For Use with MATLAB, Version 1.2, Department of Automatic Control and Systems Engineering, University of Sheffield, Sheffield, 1994.

A. R. Conn, N. I. M. Gould, P. L. Toint, Trust Region Methods, SIAM, Philadelphia, Pennsylvania, 2000.

W. R. Madych, Miscellaneous Error Bounds for Multiquadric and Related Interpolators, Computers and Mathematics with Applications 24 (12) (1992) 121–138.

M. E. Johnson, L. M. Moore, D. Ylvisaker, Minimax and Maximin Distance Designs, Journal of Statistical Planning and Inference 26 (2) (1990) 131–148.

J. Sobieszczansk-Sobieski, R. Haftka, Multidisciplinary Aerospace Design Optimization: Survey of Recent Developments, Structural Optimization 14 (1) (1997) 1–23.

M. Drela, H. Youngren, XFOIL 6.9 User Primer, Department of Aeronautics and Astronautics, Massachusetts Institute of Technology, Cambridge, MA, 2001.

Y. Tenne, K. Izui, S. Nishiwaki, A Computational Intelligence Algorithm for Simulation-driven Optimization Problems, in: Proceedings of the Third International Conference on Future Computational Technologies and Applications–Future Computing 2011, International Academy, Research, and Industry Association (IARIA), Xpert Publishing Services, 127–134, 2011b.

X. Wu, V. Kumar, R. J. Quinlan, J. Ghosh, Q. Yang, H. Motoda, G. J. McLachlan, A. Ng, B. Liu, P. S. Yu, Z.-H. Zhou, M. Steinbach, D. J. Hand, D. Steinberg, Top 10 Algorithms in Data Mining, Knowledge and Information Systems 14 (2008) 1–37.

R. M. Hicks, P. A. Henne, Wing Design by Numerical Optimization, Journal of Aircraft 15 (7) (1978) 407–412.

H.-Y. Wu, S. Yang, F. Liu, H.-M. Tsai, Comparison of Three Geometric Representations of Airfoils for Aerodynamic Optimization, in: Proceedings of the 16*th* AIAA Computational Fluid Dynamics Conference, American Institute of Aeronautics and Astronautics, AIAA 2003-4095, 2003.

A. Ratle, Optimal Sampling Strategies for Learning a Fitness Model, in: The 1999 IEEE Congress on Evolutionary Computation–CEC 1999, IEEE, Piscataway, New Jersey, 2078–2085, 1999.

N. Hansen, A. Ostermeier, Completely derandomized self-adaptation in evolution strategies, Evolutionary Computation 9 (2) (2001) 159–195, `http://www.lri.fr/~hansen/cmaes_inmatlab.html`.

D. R. Jones, M. Schonlau, W. J. Welch, Efficient Global Optimization of Expensive Black-Box Functions, Journal of Global Optimization 13 (1998) 455–492.

J. Sacks, W. J. Welch, T. J. Mitchell, H. P. Wynn, Design and Analysis of Computer Experiments, Statistical Science 4 (4) (1989) 409–435.

J. D. Martin, T. W. Simpson, Use of Kriging Models to Approximate Deterministic Computer Models, AIAA Journal 43 (4) (2005) 853–863.

S. N. Lophaven, H. B. Nielsen, J. Søndergaard, Aspects of the MATLAB Toolbox DACE, Technical report IMM-REP-2002-13, Technical University of Denmark, 2002.

R. Hooke, T. Jeeves, Direct Search Solution of Numerical and Statistical Problems, Journal of the ACM 8 (2) (1961) 212–229.

In: Global Optimization
Editor: Angelika Michalski
ISBN: 978-1-62417-796-5

Chapter 6

CUTTING BOX STRATEGY: AN ALGORITHMIC FRAMEWORK FOR IMPROVING METAHEURISTICS FOR CONTINUOUS GLOBAL OPTIMIZATION

***Wendel Melo*[1,2,*], *Marcia Fampa*[1,2] *and Fernanda Raupp*[3]**
[1]Institute of Mathematics, Federal University of Rio de Janeiro, Brazil
[2]Alberto Luiz Coimbra Institute - Graduate school and Research in Engineering (COPPE), Federal University of Rio de Janeiro, Brazil
[3]Department of Industrial Engineering, Pontifical Catholic University of Rio de Janeiro, Brazil

Abstract

In this work, we present a new framework to increase effectiveness of metaheuristics in seeking good solutions for the general nonlinear optimization problem, called *Cutting Box Strategy* (CBS). CBS is based on progressive reduction of the search space through the use of intelligent multi-starts, where solutions already obtained cannot be revisited by the adopted metaheuristic. Computational experiments with the CBS strategy are conducted with a variant of the population-based metaheuristic Differential Evolution to solve 36 test instances. The numerical results show that CBS can substantially increase the quality of the results of a metaheuristic applied for a nonlinear optimization problems.

*E-mail address: wendelmelo@cos.ufrj.br

Keywords: Metaheuristic, Constrained Optimization, Continuous Global Optimization, Nonlinear Programming

1. Introduction

Continuous global optimization problems arise in several practical applications related to natural, exact and economic sciences. In a general form, the problem can be express as finding a $x^* \in \mathbb{R}^n$ that solves:

$$\begin{aligned} \text{minimize}_x \quad & f(x) & & (1)\\ \text{subject to} \quad & g_i(x) \leq 0, & i = 1, \ldots, p, \quad & (2)\\ & h_j(x) = 0, & j = 1, \ldots, q, \quad & (3)\\ & l_k \leq x_k \leq u_k, & k = 1, \ldots, n, \quad & (4) \end{aligned}$$

where f, g_i, $i = 1, \ldots, p$, and h_j, $j = 1, \ldots, q$ are real-valued functions, not necessarily convex neither continuous nor differentiable, and l and u are n-dimensional vectors denoting lower and upper bounds to variables, respectively, used to define the *box constraints* (4). Denoting the feasible region of problem (1) by $\mathcal{F}$, we say that $\bar{x} \in \mathcal{F}$ is a local optimal solution if $f(\bar{x}) \leq f(x)$ for all $x \in \mathcal{F}$ in a neighborhood of $\bar{x}$. A solution $x^* \in \mathcal{F}$ such that $f(x^*) \leq f(x)$ for all $x \in \mathcal{F}$ is denoted global optimal solution, which is the solution of interest.

Classical continuous methods that address problem (1) may have trouble in finding some of their global optimal solutions, converging to a local optimal solution, or even to a solution that is neither a local nor a global optimal. In some cases, these methods cannot achieve a feasible solution, even though the feasible region is non-empty. This difficulty in finding a global optimal solution to the addressed problem can become critical if the problem (1) is not convex, or any of its objective or constraint functions is not continuous or not differentiable at some point in $\{x^* \in \mathbb{R}^n \mid l \leq x \leq u\}$.

To circumvent these difficulties, many researchers have developed studies on the application of stochastic local search procedures or metaheuristics that solve problem (1) (see, for example, [1, 2, 4, 10–13, 15, 17–19, 21–23, 25, 27–29, 32]). Metaheuristics are nondeterministic procedures that make use of randomness in a variety of strategies in order to avoid premature convergence to local optimal solutions. Thereby, it is expected that when applying a metaheuristic to solve problem (1) we would obtain a global optimal solution, or at least a solution close to it. The quality of the solution is, in many cases, closely related to the efficacy of the algorithm in exploring the search region in a diverse way and escaping from any local optima possibly found.

Differential Evolution (DE) is one of the most successful metaheuristic approaches for continuous optimization problems [5, 6, 24, 26, 31]. Its general idea is to use a population of candidate solutions that evolve over an iterative process searching for good solutions with the application of crossover and mutation operations. Proposed by Storn and Price in [30], the original algorithm was developed for box constrained problems (i.e., optimization problems with box constraints only). Then, many researches extended its application to solve problem (1) using a variety of strategies [2, 11, 13, 15, 17, 21, 23, 27, 32, 33]. For example, in [17], Melo et al. propose the Differential Evolution Variant (DEV), that adopts advanced techniques to solve problem (1) more effectively, such as adaptive parameters, dynamic update of the population, multiple offspring and a stopping criterion that detects the convergence of the population to a feasible solution of the problem, which is an alternative to the maximum number of iterations, originally proposed.

In this work, we propose a new mechanism, named *Cutting Box Strategy*, to increase the efficiency of local search procedures and metaheuristics on the solution of problem (1), especially when there are multiple local optima. We incorporate this strategy to the already proposed algorithm DEV and evaluate the results for a set of benchmark instances. Although in this work we only evaluate the use of the Cutting Box Strategy for DEV, this strategy is independent from the DE algorithm and can be combined with any other population-based metaheuristic or stochastic local search algorithm that has some mechanism for detecting convergence.

This chapter is organized as follows: Section 2 describes the DEV algorithm, whereas Section 3 presents the Cutting Box Strategy. Computational results are shown in Section 4. Finally, Section 5 presents the conclusions of this work.

2. Differential Evolution Variant

2.1. Differential Evolution

In [30], Storn and Price proposed the Differential Evolution (DE) method, a population-based metaheuristic for continuous optimization problems considering only box constraints. The main idea of this approach is to use a population of candidate solutions that evolve over an iterative process to find the best possible value for the objective function. Given the current population at iteration g, a new descendant solution is generated for each candidate solution of this population. If the descendant solution proves to be better than its respective ascendent solution (target solution), then it will take the place of the former. Otherwise, the descendant solution is discarded and the target solution remains in the population.

In the original DE algorithm, only a single descendant solution is generated for each candidate solution of the current population at each iteration (or generation). However, some DE competitive approaches (e.g., [23, 32]) consider multiple descendant solutions, which consists in generating M descendant solutions for each candidate solution of the population at each generation. If the best solution among the M descendant solutions is better than its respective target solution, then it will take the place of the former generated solution in the population (in this situation, the other $M - 1$ descendant solutions are discarded along with the target solution). Otherwise, the target solution remains in the population and the M descendant solutions are discarded. Algorithm 1 shows DE with multiple offspring based on the variation of DE known as DE rand/1/bin [30]. Note that all solutions of the population are used, each in turn, as a target solution. For generating new candidate solutions this approach employs crossover and mutation operators, lines 9-12, where three other random solutions besides the target solution are used. Although the offspring are generated from four solutions, they compete for entry into the population only with the target solution. In general, when a coordinate of a generated descendant solution does not satisfy the box constraint (4), a new random value is generated between its limits. As shown in Algorithm 1, only the maximum number of iterations ($MAXGEN$) is used as a stopping criterion, likewise the original DE approach.

2.2. Differential Evolution Variant Aspects

Since the original DE algorithm was developed to minimize functions subject to box constraints, the criterion presented to select the best solution among the target solution and its descendants in line 13 of Algorithm 1, is to choose the solution with the lowest value for the objective function. The DE approaches that address problem (1), however, must adopt more sophisticated selection cri-

Input: *MAXGEN*: number of generations, P: size of population, M: number of descendants, *CR*: crossover parameter.

Generate and evaluate an initial population $x^{i,0}, i = 1, \ldots, P$;
Set F as a real value between 0 and 1 ;
for $g = 1, \ldots, MAXGEN$ **do**
    for $k = 1, \ldots, P$ **do**
        for $i = 1, \ldots, M$ **do**
            Randomly select $r_1 \neq r_2 \neq r_3 \neq k \in \{1 \ldots P\}$;
            $r_{nbr} \leftarrow randint(1, n)$;
            for $j = 1, \ldots n$ **do**
                if $rand(0, 1) \leq CR$ **or** $j = r_{nbr}$ **then**
                    $d_j^{k,i} \leftarrow x_j^{r_3,g} + F(x_j^{r_1,g} - x_j^{r_2,g})$;
                else
                    $d_j^{k,i} \leftarrow x_j^{k,g}$;
        Let u^k be the best solution (according to a predefined criterion) among $x^{k,g}$ and $d^{k,i}, i = 1, \ldots, M$;
        $x^{k,g+1} \leftarrow u^k$;

Algorithm 1: DE standard algorithm with multiple descendants.

teria to deal with the feasibility of the solutions (the interested reader can find a good discussion in [20]). The Differential Evolution Variant (DEV) algorithm, which is used as the basic approach in this work, has as part of its selection criteria all three Deb's comparison criteria [7], listed as follows:
(1) any feasible solution is better than any infeasible one;
(2) between two feasible solutions, the solution with lower objective function value is preferred;
(3) between two infeasible solutions, the solution with lower penalty term is better.
The Deb criteria were also used in many other works, such as, for example, [17–19, 23, 29, 32].

Algorithm 2 presents the DEV algorithm. We observe that this algorithm adopts multiple offspring and therefore uses a criterion to select one solution among each target solution and its M descendants in every generation (line 15). Based on the method proposed in [19], the selection procedure is presented in Algorithm 3. In this procedure, the parameter S_r defines a probability for the best descendant solution (according to Deb's comparison criteria) to be compared to the target solution only with respect to the objective function, as in the original DE algorithm. Thereby, with probability $1 - S_r$ these two solutions will also be compared according to the Deb's criteria. The algorithm can then incorporate "good infeasible solutions" that lead to promising regions within the feasible region. Values near 1.0 for this parameter can provide diversity in the exploration process, but may hinder the location of feasible solutions and the convergence of the algorithm. Therefore, unlike what is proposed in [19], DEV adopts dynamic updating for this parameter in line 14 of Algorithm 2. Notice that in early iterations, the value of S_r is close to S_r^0, and as the algorithm evolves the value assigned to S_r approaches 0.0. In this way, it is possible to properly include diversity at the beginning of the exploration process, and thereafter favor the convergence of the algorithm to a good feasible solution.

In DEV algorithm, parameter CR plays a role similar to S_r. This parameter defines the probability for each coordinate of the descendant solutions to be either equal to their corresponding mutant solution coordinate (line 11) or target solution coordinate (line 13). Intuitively, we could expect that values close to 0.5 would favor diversity in the exploration of the search space. However, empirical studies, e.g. [9, 32], show that values close to 0.9 lead to better performance of the DE algorithm. For this reason, at each iteration DEV updates CR according to the expression in line 3 of Algorithm 2, so that values close to CR^0 are considered at early iterations, and from iteration $\frac{MAXGEN}{2}$, CR is equal to 1.0.

Also note that at line 16 of Algorithm 2, the solution x^k of the population is updated and becomes immediately available for the crossover and mutation operators, still in the current iteration, unlike the original DE algorithm, where the updated solution only becomes available in the next iteration. This strategy had been used before in [3], and according to the authors, it accelerates the convergence of the DE algorithm, besides saving memory in the computational implementation.

Another important feature of the DEV algorithm is the incorporation of a stopping criterion alternative to the maximum number of iterations (rows 17-20). The test stops the evolutionary process if all the solutions in the population are feasible and belong to a hypercube with edge length $\epsilon > 0$. Stopping the algorithm by the detection of convergence, this criterion can save considerable computational effort. We emphasize that, when the population converges, it is not effective to continue the evolutionary process, since each coordinate of new generated solutions depends exclusively on the respective coordinate of the

Input: $MAXGEN$: number of generations, P: size of population, M: number of descendants, CR^0: initial crossover parameter, S_r^0: initial selection parameter, ϵ: convergence tolerance.

Generate and evaluate a initial population x^i, $i = 1, \ldots, P$;
for $g = 1, \ldots, MAXGEN$ **do**
    $CR \leftarrow \min\{(1 - CR^0)\frac{2g}{MAXGEN} + CR^0 \;,\; 1\}$;
    for $k = 1, \ldots, P$ **do**
        for $i = 1, \ldots, M$ **do**
            Randomly select $r_1 \neq r_2 \neq r_3 \neq k \in \{1 \ldots P\}$;
            $r_{nbr} \leftarrow randint(1, n)$;
            $F \leftarrow rand(0.3, 0.9)$;
            for $j = 1, \ldots, D$ **do**
                if $rand(0,1) < CR$ **or** $j = r_{nbr}$ **then**
                    $d_j^{k,i} \leftarrow x_j^{r_3} + F(x_j^{r_1} - x_j^{r_2})$;
                else
                    $d_j^{k,i} \leftarrow x_j^k$;
        $S_r \leftarrow S_r^0 * (1 - \frac{g}{MAXGEN})$;
        $u^k \leftarrow SelectionOp(S_r, x^k, d^k)$;
        $x^k \leftarrow u^k$;
    `// Alternative stopping rule`
    $\overline{\sigma}_i \leftarrow \max(x_i^1, x_i^2, \ldots, x_i^P),\ i = 1, \ldots, n$;
    $\underline{\sigma}_i \leftarrow \min(x_i^1, x_i^2, \ldots, x_i^P),\ i = 1, \ldots, n$;
    if $\|\overline{\sigma} - \underline{\sigma}\|_\infty < \epsilon$ **and** *all current population solutions are feasible* **then**
        stop the evolutionary process;

Algorithm 2: DEV Algorithm.

solutions already in the population, regardless of their quality (regarding the objective function value). This ability to detect the convergence of the population was the main motivation for developing the Cutting Box Strategy, which is detailed in the next section.

3. Cutting Box Strategy

Although good metaheuristics make use of mechanisms to favor the exploration of the search space with diversity, they may fail to find global solutions especially if the addressed problem presents many local optimal points. To in-

Input: S_r: selection parameter, x^k: target solution, d^k: descendant solutions of target solution.

```
Choose u as the best solution among the M descendant generated solutions (d^k)
(according to the Deb's comparison criteria) ;
if rand(0,1) ≤ S_r then
    if f(u) ≤ f(x^k) then
        x^s ← u;
    else
        x^s ← x^k ;
else
    if u is better than x^k (according to the Deb's comparison criteria) then
        x^s ← u;
    else
        x^s ← x^k;
return x^s ;
```

Algorithm 3: Procedure *SelectionOp*.

crease the effectiveness of metaheuristics in seeking good solutions, we present in this work a new algorithmic framework that aims to prevent premature convergence of the main procedure adopted. This algorithm, called *Cutting Box Strategy* (CBS), can be used with any metaheuristic that has some mechanism to detect convergence, as for example, DEV (lines 17-20 of the Algorithm 2), even if this convergence is neither to a global nor to a local optimal solution.

The main idea of the CBS strategy is to apply a metaheuristic to solve problem 1 several times, in a scheme that can be represented by a tree, the CBS tree (Figure 1). The metaheuristic adopted in the solution of 1 will be referred in the remaining of this paper as the *basic procedure*. At each level w of the tree, CBS adresses a subproblem obtained from (1) by replacing the box constraint (4):

$$l \leq x \leq u,$$

by:

$$l^w \leq x \leq u^w, \tag{5}$$

where l^w and u^w are such that the constraint (5) defines a subset of the feasilble set defined by (4), i.e., any solution $\bar{x}$ that satisfies (5) shall also satisfy (4). We, thereby, define the box $B^w = \{x \in \mathbb{R}^n \mid l^w \leq x \leq u^w\}$.

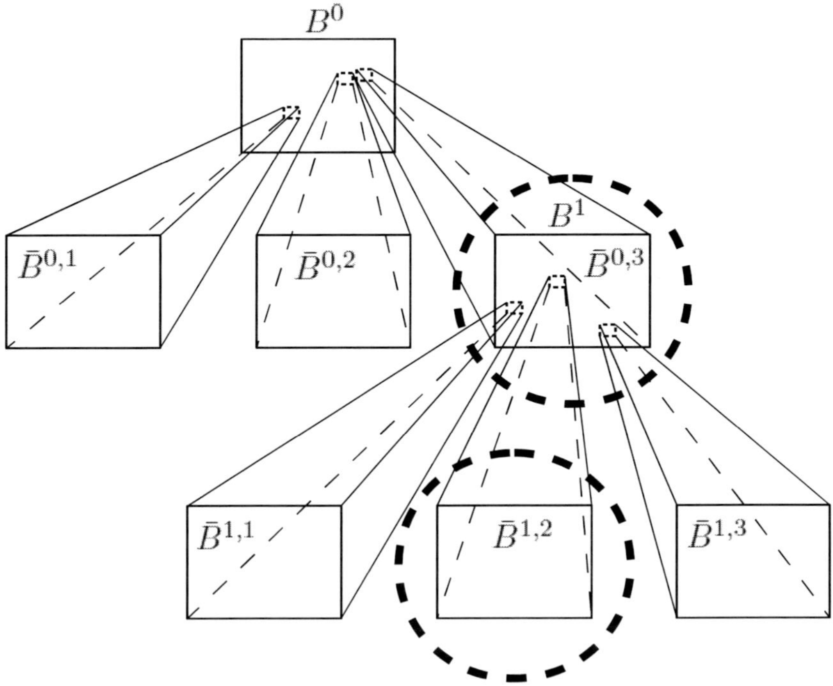

Figure 1. Example of the CBS tree with $maxSubBoxes = 3$ and $maxLevels = 2$. On level zero, box 3 ($\bar{B}^{0,3}$) presents the best solution. On level 1, box 2 ($\bar{B}^{1,2}$) presents the best solution, which is returned by the algorithm (see Algorithm 4).

Initially, at the root node of the CBS tree, $l^0 = l$ and $u^0 = u$ are set. As the algorithm CBS descends in the tree, the box that represents the search space is reduced by a factor denoted by λ and constructed around the best solution obtained on the current level. To avoid bottlenecks around wrong solutions a number of $maxSubBoxes$ calls to the basic procedure are performed on each level w. Each of these calls to the basic procedure with input B^w will result in a different solution to the addressed problem, which is accomplished by the use of the concept of *cutting box*. Let $\bar{x}^{w,z}$ be the solution obtained on the z-th call to the basic procedure on the level w. Considering the solution $\bar{x}^{w,z}$ and the reduction factor λ, define the box:

$$\bar{B}^{w,z} = \{x \in \mathbb{R}^n | l^{w,z} \leq x \leq u^{w,z}\}$$

where

$$l_i^{w,z} = \max\{\bar{x}_i^{w,z} - (u_i^w - l_i^w)\frac{\lambda}{2},\ l_i^w\},\ i = 1,\ldots,n \tag{6}$$

$$u_i^{w,z} = \min\{\bar{x}_i^{w,z} + (u_i^w - l_i^w)\frac{\lambda}{2},\ u_i^w\},\ i = 1,\ldots,n \tag{7}$$

Input: *BASICPROC*: basic procedure, $maxLevels$: maximum number of levels in a tree, $maxSubBoxes$: maximum number of boxes on each level, α: convergence factor for *BASICPROC*, λ: box reduction factor.

1 $l^0 \leftarrow l,\ u^0 \leftarrow u$;
2 let $B^w = \{x \in \mathbb{R}^n \mid l^w \le x \le u^w\}$;
3 $fbest \leftarrow \infty$;
4 **for** $w = 0,\ldots, maxLevels - 1$ **do**
5     $numSubBoxes \leftarrow 0$;
6   **for** $z = 1,\ldots, maxSubBoxes$ **do**
7   Apply *BASICPROC* to box B^w preventing solutions in boxes $\bar{B}^{w,j}$, $j = 1,\ldots,z-1$;
8   **if** *BASICPROC converges* **then**
9             $numSubBoxes \leftarrow numSubBoxes + 1$;
10   let $\bar{x}^{w,z}$ the best solution obtained in line 7 ;
11             $l_i^{w,z} \leftarrow \max\{\bar{x}_i^{w,z} - (u_i^w - l_i^w)\frac{\lambda}{2},\ l_i^w\},\ i = 1,\ldots,n;$
12             $u_i^{w,z} \leftarrow \min\{\bar{x}_i^{w,z} + (u_i^w - l_i^w)\frac{\lambda}{2},\ u_i^w\},\ i = 1,\ldots,n;$
13             $\bar{B}^{w,z} \leftarrow \{x \in \mathbb{R}^n \mid l^{w,z} \le x \le u^{w,z}\}$;
14   **else**
15   go to line 16 (*break the loop*) ;
16   **if** $numSubBoxes > 0$ **then**
17   go to line 23 (*break the loop*) ;
18     $b \leftarrow \text{argmin}_{j=1,\ldots,numSubBoxes}\{f(\bar{x}^{w,j})\}$;
19     $l^{w+1} \leftarrow l^{w,b},\ u^{w+1} \leftarrow u^{w,b}$;
20   **if** $f(\bar{x}^{w,b}) < fbest$ **then**
21         $xbest \leftarrow \bar{x}^{w,b}$;
22         $fbest \leftarrow f(\bar{x}^{w,b})$;
23 **return** $xbest$;

Algorithm 4: Cutting Box Strategy Algorithm.

In the $(z + 1)$-th call to the basic procedure on level w, this procedure adresses the problem defined over box B^w, preventing the obtained solutions to belong to any of the boxes $\bar{B}^{w,j}$, $j = 1, \ldots, z$. Regarding the DEV algorithm, this may be accomplished as follows: whenever the algorithm generates a solution in any one of these boxes, a coordinate i of this solution is randomly selected. This solution coordinate is then set equal to a random value in the range $[l_i^w, u_i^w]$. This procedure is repeated until the obtained solution no longer belongs to any of the boxes $\bar{B}^{w,j}$, $j = 1, \ldots, z$. Note that in this way, DEV works as optimizing a problem within a *cutting box*, where each new run on a certain level w generates a new cutting. Therefore, each one of the $maxSubBoxes$ runs of DEV on a given level w provides a different solution. In this way, local optima found in previous runs are discarded, increasing the possibility of the algorithm to find global optimal solutions. The best box found, i.e., the box associated to the best solution $\bar{x}^{w,z}$, is defined as box B^{w+1} on the next level of the tree. If the basic procedure fails to converge to a feasible solution on the z-th iteration of level w, the algorithm CBS advances directly to the next level considering only the boxes obtained in the previous iterations at level w. To test the convergence of the basic procedure, it is possible to use a tolerance factor α dependent on the size of B^w. For example, considering DEV as the basic procedure, we may consider the convergence tolerance $\epsilon = \alpha \max_{i=1,\ldots,n}\{u_i^w - l_i^w\}$. In cases where the ranges of the variables are disproportioned, it might be a good idea to adopt for DEV a different tolerance ϵ_i for each coordinate i, i.e., $\epsilon_i = \alpha(u_i^w - l_i^w)$.

The CBS algorithm descends in the search tree until level $maxLevels$ and then returns the best solution found. Algorithm 4 shows the CBS algorithm while Figure 1 illustrates a small example of a CBS tree.

4. Computational Results

In this section, we describe the computational results obtained with the application of the CBS framework combined with a metaheuristic to solve instances of the general nonlinear optimization problem. As previously mentioned, we adopt the DEV algorithm as the basic procedure on the application of CBS because it provides a mechanism for detecting convergence, required by CBS, and also because of its computation effectiveness, which is demonstrated in [17]. In [17], the authors report the successful application of DEV on a set of 22 test instances, which are described in [14]. In this paper, we consider a set of 36 instances with multiple local optima taken from CEC2010 (detailed information on the instances can be found in [16]). This set of test problems was constructed with 18 scalable instances, where each instance was included in the set with 10 and 30 variables, according to [16]. We note, however, that the optimal solution of these test instances is not available in the literature. As recommended in [16], we adopt a tolerance of 10^{-4} with respect to the equality constraints (3).

The experiments were conducted with our own implementations of the algorithms in C++, compiled with g++ compiler (GCC 4.4) using the flags "-O3-march = native". Tests were run on a computer with Intel Core 2 Duo frequency 2.13 GHz and 2 MB cache, 2 GB of RAM and using the Linux operating system Ubuntu 11 (32 bits). For the generation of random numbers, we used the implementation of the Mersenne Twister method available in [8] .

We compare the application of DEV itself with the application of DEV within the CBS framework, performing 100 runs on all test instances. In the first case, the parameter ϵ was set equal to $1.0e-6$, while in the second case, this parameter was defined as a function of the parameter α, as explained in Section 3. The other DEV parameters were adjusted for both cases as follows: $\mathit{MAXGEN} = 2000$, $P = 50$, $M = 4$, $\mathit{CR}^0 = 0.5$ and $Sr^0 = 1.0$ (see Algorithm 2 for a description of these parameters). The values used for the CBS parameters were $maxLevels = 2$, $maxSubBoxes = 3$, $\lambda = 0.1$ and $\alpha = 0.001$ (see Algorithm 4 for a description of these parameters). In the last level of the CBS tree, we adopted $maxSubBoxes = 1$. The values used for the parameters were determined empirically.

Tables 1 and 2 show the computational results for the 18 test instances with 10 and 30 variables, respectively. The first column (Problem) identifies the problem, while the second one (Alg) discriminates the approach. Considering 100 executions of each approach for each test instance, the remaining columns bring the value of the objective function in the best solution found (Best), the average value of the objective function (Avg), the objective function value of the worst solution found (Worse), the standard deviation (SD), the number of runs in which at least one feasible solution was found (Feas Sols), the average running time in seconds (Avg time (s)), the average number of the objective function evaluations (Avg Avals-F) and the average number of the constraint functions (2) and (3) evaluations (Avg C-Avals). The last lines in each table show the average values (considering the previous lines) for the approaches.

Table 1. Computational results for DEV and CBS + DEV ($n = 10$)

Problem	Alg	Best	Avg	Worse	SD	Feas Sols	Avg Time(s)	Avg F-Avals	Avg C-Avals
C01	DEV	-0.7473	-0.7465	-0.7259	3.87E-03	100	0.50	140,086	196,712
$n = 10$	CBS + DEV	-0.7473	-0.7473	-0.7473	1.26E-08	100	1.54	418,399	608,236
C02	DEV	-2.2777	-2.2681	-2.2308	1.08E-02	99	0.83	207,902	321,604
$n = 10$	CBS + DEV	-2.2777	-2.2775	-2.2717	1.12E-03	100	2.05	452,558	833,092
C03	DEV	5.80E-14	8.69804	8.87555	1.25E+00	100	0.09	98,441	158,054
$n = 10$	CBS + DEV	7.07E-14	8.52065	8.87555	1.75E+00	100	0.51	489,934	855,472
C04	DEV	-9.92E-06	-9.85E-06	-9.77E-06	2.91E-08	100	0.27	75,252	134,244
$n = 10$	CBS + DEV	-9.93E-06	-9.86E-06	-9.78E-06	2.78E-08	100	1.17	321,969	605,320
C05	DEV	-483.611	-483.611	-483.611	2.74E-08	100	0.41	79,040	141,494
$n = 10$	CBS + DEV	-483.611	-483.611	-483.611	2.25E-08	100	1.91	358,105	675,874
C06	DEV	-578.662	-578.662	-578.657	7.20E-04	100	2.02	258,315	392,142
$n = 10$	CBS + DEV	-578.662	-578.662	-578.661	2.02E-04	100	5.52	644,671	1,125,880
C07	DEV	3.27E-14	9.41E-14	2.56E-13	4.27E-14	100	0.26	117,171	117,976
$n = 10$	CBS + DEV	1.98E-14	8.82E-14	2.87E-13	4.25E-14	100	0.46	202,536	204,978
C08	DEV	2.38E-14	2.4498	10.9415	4.45E+00	100	0.54	103,038	124,544
$n = 10$	CBS + DEV	1.48E-14	0.3607	10.9415	1.65E+00	100	1.21	216,461	289,460
C09	DEV	2.44E-14	2.0718	4.4082	2.21E+00	100	0.26	80,040	126,700
$n = 10$	CBS + DEV	2.04E-14	1.0140	4.4082	1.86E+00	100	0.82	231,586	409,480
C10	DEV	5.30E-14	38.8051	41.7260	1.07E+01	100	0.86	126,257	208,152
$n = 10$	CBS + DEV	2.77E-14	37.1361	41.7262	1.31E+01	100	2.89	393,998	715,520
C11	DEV	-	-	-	0.00E+00	0	1.95	400,050	400,050
$n = 10$	CBS + DEV	-	-	-	0.00E+00	0	1.95	400,050	400,050
C12	DEV	-	-	-	0.00E+00	0	1.10	215,085	400,050
$n = 10$	CBS + DEV	-	-	-	0.00E+00	0	1.09	215,380	400,050
C13	DEV	-68.4294	-68.4294	-68.4294	5.82E-07	100	1.66	254,853	327,326
$n = 10$	CBS + DEV	-68.4294	-68.4294	-68.4294	3.65E-07	100	3.42	484,093	707,938
C14	DEV	2.05E-14	8.63E-14	3.83E-13	4.87E-14	100	0.54	113,913	121,352
$n = 10$	CBS + DEV	2.17E-14	8.91E-14	6.66E-13	7.25E-14	100	0.97	187,303	230,278
C15	DEV	2.27E-14	3.2692	3.6732	1.16E+00	100	0.96	103,807	153,988
$n = 10$	CBS + DEV	4.31E-14	2.8225	3.6732	1.50E+00	100	2.39	235,291	392,044

Table 1. Continued

Problem	Alg	Best	Avg	Worse	SD	Feas Sols	Avg Time(s)	Avg F-Avals	Avg C-Avals
C16	DEV	0	1.54E-15	1.31E-14	2.52E-15	100	0.34	37,815	66,390
$n = 10$	CBS + DEV	0	6.89E-16	5.90E-15	1.41E-15	100	2.14	227,495	424,288
C17	DEV	1.67E-14	0.0322	2.9011	3.06E-01	90	0.30	83,931	148,926
$n = 10$	CBS + DEV	1.17E-14	2.76E-09	2.51E-07	2.63E-08	91	1.21	328,482	615,150
C18	DEV	1.22E-25	1.73E-18	1.14E-16	1.17E-17	100	0.53	137,999	170,662
$n = 10$	CBS + DEV	4.91E-27	4.18E-18	3.24E-16	3.25E-17	100	1.68	359,987	573,032
AVG	DEV	-188.9546	-98.0355	-96.4662	1.83E+00	88	0.74	146277	206131
$n = 10$	CBS + DEV	-188.9546	-108.3873	-106.4096	2.21E+00	88	1.83	342683	525898

Table 2. Computational Results for DEV and CBS + DEV ($n = 30$)

Problem	Alg	Best	Avg	Worse	SD	Feas Sols	Avg Time(s)	Avg F-Avals	Avg C-Avals
C01	DEV	-0.8219	-0.8170	-0.7970	5.97E-03	100	1.78	198,204	248,524
$n = 30$	CBS + DEV	-0.8219	-0.8213	-0.8179	1.25E-03	100	5.52	583,333	787,746
C02	DEV	-2.1787	-1.8375	-1.2136	2.49E-01	100	2.66	242,334	382,998
$n = 30$	CBS + DEV	-2.2258	-2.1466	-1.0326	1.94E-01	100	7.24	606,947	1,081,060
C03	DEV	28.6735	28.6735	28.6735	2.86E-06	100	0.24	115,227	199,274
$n = 30$	CBS + DEV	28.6735	28.6735	28.6735	1.82E-06	100	0.88	401,346	740,778
C04	DEV	-	-	-	0.00E+00	0	2.05	208,419	400,050
$n = 30$	CBS + DEV	-	-	-	0.00E+00	0	2.05	208,419	400,050
C05	DEV	-152.873	-68.192	81.053	4.03E+01	84	3.22	236,669	398,830
$n = 30$	CBS + DEV	-231.581	-153.885	-88.324	3.18E+01	82	8.43	579,475	1,069,030
C06	DEV	-513.581	-268.165	-91.7801	1.29E+02	99	6.00	252,325	398,278
$n = 30$	CBS + DEV	-528.739	-392.73	-194.225	1.16E+02	99	15.38	590,340	1,082,330
C07	DEV	6.7740	15.1399	72.6330	1.42E+01	100	1.31	251,400	254,402
$n = 30$	CBS + DEV	5.4182	9.7708	15.4738	2.09E+00	100	2.54	478,792	487,884
C08	DEV	10.0434	21.076	118.686	2.17E+01	100	3.29	223,933	255,610
$n = 30$	CBS + DEV	6.75828	10.904	15.8365	1.82E+00	100	7.50	468,721	597,180

Table 2. Continued

Problem	Alg	Best	Avg	Worse	SD	Feas Sols	Avg Time(s)	Avg F-Avals	Avg C-Avals
C09	DEV	481.97	4,810.52	26,692.00	5.11E+03	100	2.06	255,335	368,486
$n = 30$	CBS + DEV	97.81	268.36	2,937.10	3.53E+02	100	5.27	562,196	984,846
C10	DEV	251.37	2,739.53	25,440.50	3.22E+03	100	4.73	258,340	369,554
$n = 30$	CBS + DEV	53.63	180.94	1,771.81	2.87E+02	100	12.01	567,108	979,536
C11	DEV	-	-	-	0.00E+00	0	5.90	400,050	400,050
$n = 30$	CBS + DEV	-	-	-	0.00E+00	0	6.00	400,050	400,050
C12	DEV	-	-	-	0.00E+00	0	3.04	208,409	400,050
$n = 30$	CBS + DEV	-	-	-	0.00E+00	0	3.04	208,455	400,050
C13	DEV	-66.9411	-63.1009	-57.0551	1.88E+00	100	5.76	310,418	400,050
$n = 30$	CBS + DEV	-67.4158	-64.7474	-62.8394	1.04E+00	100	13.40	658,411	985,986
C14	DEV	11.9862	18.635	75.201	1.16E+01	100	3.14	223,475	254,376
$n = 30$	CBS + DEV	8.18846	22.0326	85.6042	2.44E+01	100	6.80	441,256	577,406
C15	DEV	21.9435	59.0175	454.892	6.22E+01	100	4.87	183,081	265,536
$n = 30$	CBS + DEV	21.6046	21.728	27.3195	8.03E-01	100	14.51	491,955	837,683
C16	DEV	7.29E-14	0.0001	0.0132982	1.33E-03	100	3.13	118,935	211,110
$n = 30$	CBS + DEV	1.68E-15	1.17E-14	2.80E-14	4.78E-15	100	10.85	395,967	742,782
C17	DEV	0.0706	0.5871	3.1105	7.09E-01	100	1.87	216,135	324,316
$n = 30$	CBS + DEV	0.0127	0.0626	1.0916	1.11E-01	100	5.21	532,203	948,396
C18	DEV	3.7411	19.2012	77.9627	1.27E+01	100	2.92	240,601	341,366
$n = 30$	CBS + DEV	0.0360	1.9193	6.5721	1.44E+00	100	7.36	531,697	920,918
AVG	DEV	5.7265	487.3512	3526.2586	5.08E+02	82	3.2202	230,183	326,270
$n = 30$	CBS + DEV	-43.4749	-4.9962	324.4459	51.2059	82	7.4442	483,704	719,419

Regarding the instances with 10 variables (Table 1), we note that CBS shows better results than DEV for 8 out off the 18 instances, considering the average value of the objective function (C01, C02, C03, C08, C09, C10, C15 and C17). For the other instances, the performance of both approaches can be considered equivalent with respect to this criterion. CBS also shows the best results with respect to the worst solution for 3 instances (C01, C02 and C17). Regarding the best solution, both approaches have equivalent performence on all test instances. We note that, considering these three criteria, CBS does not underperform DEV on any instance. For instances C11 and C12, no approach can find any feasible solution. We point out that, since CBS uses DEV as the basic procedure in this study, in cases where DEV fails to find a feasible solution, CBS also fails. As would be expected, the computational effort demanded by CBS was greater than that demanded by DEV. On average, the computational time for CBS is 2.45 times the time for DEV. We emphasize that each run of DEV within the CBS framework tends to be computationally cheaper than an independent run of DEV due to difference in the value of the convergence tolerance ϵ that was used. The number of evaluations of the objective function and the constraints performed by CBS are, on average, 2.34 and 2.55 times higher than the ones performed by DEV, respectively.

Regarding the instances with 30 variables (Table 2), we note that CBS shows better results than DEV for 12 out off the 18 instances (all except C1, C3, C4, C11, C12 and C16), considering the best obtained solution. For other instances, both approaches show equivalent performance with respect to this criterion. Considering the average value of the objective function, CBS shows better performance than DEV for 13 instances (all, except C3, C4, C11, C12 and C14). Only for instance C14, DEV shows better result than CBS, considering the average value of the objective function. This same behavior is repeated for the worst solution, where CBS gives better results than DEV for 12 instances (all except C02, C03, C04, C11, C12, and C14). With respect to the computational time and the average number of evaluations of the objective and constraints functions, the results for CBS are on average 2.31, 2.10 and 2.2 higher than those for DEV, respectively. Finally, we point out that both approaches cannot provide any feasible solution for instances C04, C11 and C12.

The computational results presented in this section, especially those on Table 2 demonstrate the effectiveness of the CBS algorithm. It is clear from the results that the application of CBS can improve the performance of the metaheuristic adopted to solve the problem addressed here, although this metaheuristic is applied multiple times on the same problem. The main reasons for the success of CBS are to consider boxes that gradually become smaller through the levels of the CBS tree, and its ability to prevent local optima already found to be revisited in future applications of the basic procedure on the same level of CBS tree.

5. Conclusion

Continuous global optimization problems are in general difficult to adress with classical solution methods. Although in recent years there has been a major effort in developing good metaheuristics that attempt to reach the global optimal solution of these problems, in many cases these algorithms still fail to accomplish this goal, especially when the problem has many local optima.

In this paper, we propose a new framework to improve the performance of metaheuristics for the general nonlinear optimization problem called Cutting Box Strategy (CBS). CBS uses a basic procedure that is successively applied in an optimization process inside a box, where the boxes are progressively reduced. CBS can be considered as a search tree algorithm, where the search spaces are boxes. On each level of the tree the basic procedure is applied several times to the problem, considering only smaller boxes that are contained in the original one. The different aspect of CBS is that it relies on an intelligent application of multi-starts to the subproblems on each level of the tree, which avoids the same solution to be generated more than once, by creating a "hole" in the box where the search for solutions is performed, and consequently preventing the basic procedure to generate solutions in these "holes".

The methodology described here is general and therefore can be applied to any stochastic basic procedure or metaheuristic with a mechanism for detecting convergence that can handle the addressed problem. We evaluated the performance of CBS combined with a good metaheuristic on a set of 36 test instances with multiple local optima. The results showed that CBS was able to considerably improve the results of the metaheuristic adopted as the basic procedure considering the quality of the solutions obtained in exchange for a tolerable increase in computational effort.

As possible future research, we suggest the evaluation of the performance of CBS with other basic procedures and test instances, along with the development of other strategies for exploiting the CBS tree beyond the search procedure used here.

References

[1] Shamim Akhtar, Kang Tai, and Tapabrata Ray. A socio-behavioural simulation model for engineering design optimization. In *Projects in India, by Ruth Alsop with Ved*, pages 200–2, 2002.

[2] M.M. Ali and Z. Kajee-Bagdadi. A local exploration-based differential evolution algorithm for constrained global optimization. *Applied Mathematics and Computation*, 208(1):31 – 48, 2009.

[3] B.V. Babu and Rakesh Angira. Modified differential evolution (mde) for optimization of non-linear chemical processes. *Computers & Chemical Engineering*, 30:989 – 1002, 2006.

[4] S. Ben Hamida and M. Schoenauer. Aschea: new results using adaptive segregational constraint handling. In *Evolutionary Computation, 2002. CEC '02. Proceedings of the 2002 Congress on*, volume 1, pages 884 – 889, May 2002.

[5] Uday K. Chakraborty. *Advances in Differential Evolution*. Springer Publishing Company, Incorporated, 1 edition, 2008.

[6] S. Das and P.N. Suganthan. Differential evolution: A survey of the state-of-the-art. *Evolutionary Computation, IEEE Transactions on*, 15(1):4 –31, Feb. 2011.

[7] Kalyanmoy Deb. An efficient constraint handling method for genetic algorithms. *Computer Methods in Applied Mechanics and Engineering*, 186:311 – 338, 2000.

[8] A. Fog. Pseudo random number generators uniform and non-uniform distributions. http://www.agner.org/random/ accessed 7-July-2012.

[9] Roger Gämperle, Sibylle D. Müller, and Petros Koumoutsakos. A parameter study for differential evolution. In *WSEAS Int. Conf. on Advances in Intelligent Systems, Fuzzy Systems, Evolutionary Computation*, pages 293–298. Press, 2002.

[10] Abdel-Rahman Hedar and Masao Fukushima. Derivative-free filter simulated annealing method for constrained continuous global optimization. *J. of Global Optimization*, 35(4):521–549, August 2006.

[11] F. Huang, Ling Wang, and Qie He.

[12] Slawomir Koziel and Zbigniew Michalewicz. Evolutionary algorithms, homomorphous mappings, and constrained parameter optimization. *Evol. Comput.*, 7(1):19–44, March 1999.

[13] J. Lampinen. A constraint handling approach for the differential evolution algorithm. In *Evolutionary Computation, 2002. CEC '02. Proceedings of the 2002 Congress on*, volume 2, pages 1468 –1473, 2002.

[14] J.J. Liang, T.P. Runarsson, E. Mezura-Montes, M. Clerc, P.N. Suganthan, C.A. Coello Coello, and K. Deb. Problem definitions and evaluation criteria for the cec 2006 special session on constrained real-parameter optimization. In: Report, 2006.

[15] Hui Liu, Zixing Cai, and Yong Wang. Hybridizing particle swarm optimization with differential evolution for constrained numerical and engineering optimization. *Applied Soft Computing*, 10(2):629 – 640, 2010.

[16] R. Mallipeddi and P. N. Suganthan. Problem definitions and evaluation criteria for the cec 2010 competition on constrained real- parameter optimization. In: Report, 2010.

[17] Wendel Melo, Marcia Fampa, and Fernanda Raupp. Differential evolution variant for constrained global continuous optimization. Submitted to *Computers & Industrial Engineering*, 2012.

[18] Wendel Melo, Marcia Fampa, and Fernanda Raupp. A stochastic local search algorithm for constrained continuous global optimization. *International Transactions in Operational Research*, 2012.

[19] E. Mezura-Montes and C.A.C. Coello. A simple multimembered evolution strategy to solve constrained optimization problems. *Evolutionary Computation, IEEE Transactions on*, 9(1):1 – 17, Feb. 2005.

[20] Efén Mezura-Montes and Carlos A. Coello Coello. Constraint-handling in nature-inspired numerical optimization: Past, present and future. *Swarm and Evolutionary Computation*, 1(4):173 – 194, 2011.

[21] Efrén Mezura-Montes, Carlos A. Coello Coello, Edy I. Tun-Morales, Sección De Computación, and Villahermosa Tabasco. Simple feasibility rules

and differential evolution for constrained optimization. In *In Proceedings of the 3rd Mexican International Conference on Artificial Intelligence (MI-CAI 2004*, pages 707–716. Springer Verlag, 2004.

[22] Efrén Mezura-Montes, Carlos A. Coello Coello, and Jesús Velázquez-Reyes. Increasing successful offspring and diversity in differential evolution for engineering design. In *Proceedings of the Seventh International Conference on Adaptive Computing in Design and Manufacture*, ACDM 2006, pages 131–139, 2006.

[23] Efrén Mezura-Montes, Jesús Velázquez-Reyes, and Carlos A. Coello Coello. Promising infeasibility and multiple offspring incorporated to differential evolution for constrained optimization. In *Proceedings of the 2005 conference on Genetic and evolutionary computation*, GECCO '05, pages 225–232, New York, NY, USA, 2005. ACM.

[24] Ferrante Neri and Ville Tirronen. Recent advances in differential evolution: a survey and experimental analysis. *Artificial Intelligence Review*, 33:61–106, 2010. 10.1007/s10462-009-9137-2.

[25] Chandra Sekhar Pedamallu and Linet Ozdamar. Investigating a hybrid simulated annealing and local search algorithm for constrained optimization. *European Journal of Operational Research*, 185(3):1230 – 1245, 2008.

[26] Kenneth Price, Rainer M. Storn, and Jouni A. Lampinen. *Differential Evolution: A Practical Approach to Global Optimization (Natural Computing Series)*. Springer-Verlag New York, Inc., Secaucus, NJ, USA, 2005.

[27] A.K. Qin, V.L. Huang, and P.N. Suganthan. Differential evolution algorithm with strategy adaptation for global numerical optimization. *Evolutionary Computation, IEEE Transactions on*, 13(2):398 –417, April 2009.

[28] T. Ray and K.M. Liew. Society and civilization: An optimization algorithm based on the simulation of social behavior. *Evolutionary Computation, IEEE Transactions on*, 7(4):386 – 396, Aug. 2003.

[29] T.P. Runarsson and Xin Yao. Search biases in constrained evolutionary optimization. *Systems, Man, and Cybernetics, Part C: Applications and Reviews, IEEE Transactions on*, 35(2):233 –243, May 2005.

[30] Rainer Storn and Kenneth Price. Differential evolution a simple and efficient heuristic for global optimization over continuous spaces. *J. of Global Optimization*, 11(4):341–359, December 1997.

[31] Zhenyu Yang, Xin Yao, and Jingsong He. Making a difference to differential evolution. In Patrick Siarry and Zbigniew Michalewicz, editors, *Advances in Metaheuristics for Hard Optimization*, Natural Computing Series, pages 397–414. Springer Berlin Heidelberg, 2007.

[32] Min Zhang, Wenjian Luo, and Xufa Wang. Differential evolution with dynamic stochastic selection for constrained optimization. *Information Sciences*, 178(15):3043 – 3074, 2008. <ce:title>Nature Inspired Problem-Solving</ce:title>.

[33] Karin Zielinski and Rainer Laur. Stopping criteria for differential evolution in constrained single-objective optimization. In Uday Chakraborty, editor, *Advances in Differential Evolution*, volume 143 of *Studies in Computational Intelligence*, pages 111–138. Springer Berlin / Heidelberg, 2008. 10.1007/978-3-540-68830-3_4.

INDEX

A

B

C

D

E

F

G

H

I

J

K

L

M

N

O

P